Decision Analysis for Management Judgment

Decision Analysis for Management Judgment

Paul Goodwin
Bristol Business School
George Wright
Strathclyde Graduate
Business School

With a Foreword by **Lawrence D. Phillips**

JOHN WILEY & SONS
Chichester · New York · Brisbane · Toronto · Singapore

Other Wiley Editorial Offices

John Wiley & Sons, Inc., 605 Third Avenue,
New York, NY 10158-0012, USA

Jacaranda Wiley Ltd, 33 Park Road, Milton,
Queensland 4064, Australia

John Wiley & Sons (Canada) Ltd, 22 Worcester Road,
Rexdale, Ontario M9W 1L1, Canada

John Wiley & Sons (SEA) Pte Ltd, 37 Jalan Pemimpin 05-04,
Block B, Union Industrial Building, Singapore 2057

Library of Congress Cataloging-in-Publication Data:

Goodwin, Paul.
    Decision analysis for management judgment / Paul Goodwin,
George Wright ; with a foreword by Lawrence D. Phillips.
        p.   cm.
    Includes bibliographical references and index.
    ISBN 0-471-92833-X (ppc)
    1. Decision-making.   I. Wright, George, 1952–   .   II. Title.
HD30.23.G66   1991
658.4'03—dc20   90–28576
                                                                    CIP

British Library Cataloguing in Publication Data:

A catalogue record for this book is available from the British Library.

ISBN 0-471-92833-X

Phototypeset in 10/12 pt Palatino by Dobbie Typesetting Limited, Tavistock, Devon
Printed and bound in Great Britain by Bookcraft (Bath) Ltd

To
**Kuljeet
and
Josephine, Jamie and Jerome**

Contents

Foreword   *Lawrence D. Phillips*   xiii

Preface   xv

Acknowledgments   xvii

CHAPTER 1   Introduction   1
Complex Decisions   1
The Role of Decision Analysis   3
Overview of the Book   5

CHAPTER 2   Decisions Involving Multiple Objectives   7
Introduction   7
Basic Terminology   8
   Objectives and Attributes   8
   Value and Utility   9
An Office Location Problem   9
An Overview of the Analysis   10
Constructing a Value Tree   11
Measuring How Well the Options perform on Each Attribute   13
   Direct Rating   14
   Value Functions   16
Determining the Weights of the Attributes   19
Aggregating the Benefits Using the Additive Model   21
Trading Benefits against Costs   22
Sensitivity Analysis   25
Theoretical Considerations   26
   The Axioms of the Method   26
   Assumptions Made when Aggregating Values   27
Conflicts between Intuitive and Analytic Results   29
Summary   31
Exercises   32
References   35

CHAPTER 3   Introduction to Probability   37
   Introduction   37
   Outcomes and Events   38
   Approaches to Probability   39
      The Classical Approach   39
      The Relative Frequency Approach   39
      The Subjective Approach   40
   Mutually Exclusive and Exhaustive Events   41
   The Addition Rule   42
   Complementary Events   44
   Marginal and Conditional Probabilities   44
   Independent and Dependent Events   45
   The Multiplication Rule   46
   Probability Trees   47
   Probability Distributions   48
   Expected Values   52
   The Axioms of Probability Theory   54
   Summary   54
   Exercises   55
   References   59

CHAPTER 4   Decision Making under Uncertainty   60
   Introduction   60
   The Expected Monetary Value (EMV) Criterion   61
   Limitations of the EMV Criterion   63
   Single-attribute Utility   65
   Interpreting Utility Functions   72
   Utility Functions for Non-monetary Attributes   74
   The Axioms of Utility   76
   More on Utility Elicitation   79
   How Useful is Utility in Practice?   82
   Multi-attribute Utility   86
      The Decanal Engineering Corporation   86
      Mutual Utility Independence   87
      Deriving the Multi-attribute Utility Function   88
      Interpreting Multi-attribute Utilities   93
      Further Points on Multi-attribute Utility   94
   Summary   94
   Exercises   95
   References   101

CHAPTER 5   Decision Trees and Influence Diagrams   102
   Introduction   102
   Constructing a Decision Tree   103

Determining the Optimal Policy   105
Decision Trees and Utility   107
Decision Trees Involving Continuous Probability Distributions   109
Practical Applications of Decision Trees   111
Assessment of Decision Structure   112
Eliciting Decision Tree Representations   118
Summary   122
Exercises   122
References   127

CHAPTER 6   Applying Simulation to Decision Problems   129
Introduction   129
Monte Carlo Simulation   130
Applying Simulation to a Decision Problem   133
   The Elite Pottery Company   133
   Plotting the Two Distributions   140
   Determining the Option with the Highest Expected Utility   140
   Stochastic Dominance   141
   The Mean–Standard Deviation Approach   143
Applying Simulation to Investment Decisions   146
   The Net Present Value (NPV) Method   146
   Using Simulation   149
   Utility and Net Present Value   151
Modeling Dependence Relationships   152
   Judgmental Problems   153
   Simulating Dependence Relationships   154
Summary   157
Exercises   158

CHAPTER 6–APPENDIX   The Standard Deviation   163
Example   163
References   164

**CHAPTER 7   Revising Judgments in the Light of
         New Information   165**
Introduction   165
Bayes' Theorem   166
   Example   169
   Answer   169
   Another Example   170
The Effect of New Information on the Revision of
  Probability Judgments   171
Applying Bayes' Theorem to a Decision Problem   174

Assessing the Value of New Information 175
 The Expected Value of Perfect Information 177
 The Expected Value of Imperfect Information 179
Practical Considerations 184
Summary 186
Exercises 186
References 191

CHAPTER 8 The Quality of Human Judgment: Laboratory
     Studies 192
Introduction 192
Heuristics in Probability Assessment 193
Revision of Probabilistic Opinion 198
Individual and Situational Influences on Decision Making 201
 Personality and Decision Making 201
 Cognitive Style and Decision Making 203
Contingent Decision Behavior 207
Summary 209
Discussion Questions 209
References 209

CHAPTER 9 The Quality of Human Judgment: Real-world
     Studies 212
Introduction 212
Assessing Probabilities for Future Events 212
Statistical Extrapolation 216
Econometrics 217
Judgment in Forecasting 218
Summary 221
Discussion Questions 222
References 222

CHAPTER 10 Probability Assessment 224
Introduction 224
Preparing for Probability Assessment 224
 Motivating 225
 Structuring 225
 Conditioning 225
Assessment Methods 226
 Assessment Methods for Individual Probabilities 226
 Assessment Methods for Probability Distributions 228
A Comparison of the Assessment Methods 230
Consistency and Coherence Checks 231

Assessment of the Validity of Probability Forecasts   233
Assessing Probabilities for Very Rare Events   235
   Event Trees   236
   Fault Trees   236
   Using a Log-odds Scale   239
Summary   239
Exercises   239
References   240

CHAPTER 11   Decisions Involving Groups of Individuals   242
Introduction   242
Mathematical Aggregation   243
Aggregating Judgments in General   244
   Taking a Simple Average of the Individual Judgments   245
   Taking a Weighted Average of the Individual Judgments   245
Aggregating Probability Judgments   246
Aggregating Preference Judgments   248
   Aggregating Preference Orderings   248
   Aggregating Values and Utilities   250
Unstructured Group Processes   252
Structured Group Processes   253
Decision Conferencing   255
Summary   257
Discussion Questions   257
References   258

CHAPTER 12   Resource Allocation and Negotiation Problems   259
Introduction   259
Modeling Resource Allocation Problems   260
   An Illustrative Problem   260
   Determining the Variables, Resources and Benefits   261
   Identifying the Possible Strategies for Each Region   262
   Assessing the Costs and Benefits of Each Strategy   263
   Measuring Each Benefit on a Common Scale   265
   Comparing the Relative Importance of the Benefits   267
   Identifying the Costs and Benefits of the Packages   269
   Sensitivity Analysis   273
Summary of the Main Stages of the Analysis   274
Negotiation Models   274
An Illustrative Problem   275
Practical Applications   278
Summary   279

Discussion Questions   279
References   280

CHAPTER 13   Alternative Decision-support Systems   **281**
Introduction   281
Expert Systems   281
   What is an Expert System?   281
   What is Expert Knowledge?   283
   How is Expert Knowledge Represented in Expert Systems?   284
   An Example of an Expert System Application in Life
      Underwriting   287
Statistical Models of Judgment   289
   Recent Research   292
Comparisons   294
Summary   296
References   296

Suggested Answers to Selected Questions   **298**

Suppliers of Computer Software   **302**

Index   **303**

Foreword

It is a curious fact that although ability to take decisions is at the top of most senior executives' list of attributes for success in management, those same people are usually unwilling to spend any time developing this quality. Perhaps decision making is considered as fundamental as breathing: essential for life, a natural and automatic process. Therefore, why study it?

In this book, Paul Goodwin and George Wright show why: because research over the past 30 years has revealed numerous ways in which the process of making decisions goes wrong, usually without our knowing it. But the main thrust of this book is to show how decision analysis can be applied so that decisions are made correctly. The beauty of the book is in providing numerous decision analysis techniques in a form that makes them usable by busy managers and administrators.

Ever since decision theory was introduced in 1960 by Howard Raiffa and Robert Schlaifer of Harvard University's Business School, a succession of textbooks has chronicled the development of this abstract mathematical discipline to a potentially useful technology known as decision analysis, through to numerous successful applications in commerce, industry, government, the military and medicine. But all these books have been either inaccessible to managers and administrators or restricted to only a narrow conception of decision analysis, such as decision trees.

Unusually, this book does not even start with decision trees. My experience as a practicing decision analyst shows that problems with multiple objectives are a frequent source of difficulty in both public and private sectors: one course of action is better in some respects, but another is better on other criteria. Which to choose? The authors begin, in Chapter 2, with such a problem, and present a straightforward technology, called SMART, to handle it.

My advice to the reader is to stop after Chapter 2 and apply SMART on a problem actually bothering you. Decision analysis works best on real problems, and it is most useful when you get a result you did not expect. Sleep on it, then go back and work it through again, altering and changing your representation of the problem, or your views of it, as necessary. After several tries, you will almost certainly have deepened your understanding of the issues, and now feel comfortable with taking a decision.

If you are then willing to invest some time and effort trying out the various approaches covered in the book, the rewards should be worth it. No mathematical skills are needed beyond an ability to use a calculator to add, multiply and occasionally divide. But a willingness to express your judgments in numerical form is required (even if you are not convinced at the start), and patience in following a step-by-step process will help.

Whether your current problem is to evaluate options when objectives conflict, to make a choice as you face considerable uncertainty about the future, to assess the uncertainty associated with some future event, to decide on seeking new information before making a choice, to obtain better information from a group of colleagues, to re-allocate limited resources for more effectiveness, or to negotiate with another party, you will find sound, practical help in these pages. Even if you do not overtly apply any of the procedures in this book, the perspectives on decision making provided by decision analysis should help you to deal with complex issues more effectively and sharpen your everyday decision-making skills.

Lawrence D. Phillips
Decision Analysis Unit
London School of Economics and Political Science

Preface

In an increasingly complex world, decision analysis has a major role to play in helping decision makers to gain a greater understanding of the problems they face. The main aim of this book is to make decision analysis accessible to its largest group of potential users: managers and administrators in business and public sector organizations, most of whom, although expert at their work, are not mathematicians or statisticians. We have therefore endeavored to write a book which makes the methodology of decision analysis as 'transparent' as possible so that little has to be 'taken on trust', while at the same time making the minimum use of mathematical symbols and concepts. A chapter introducing the ideas of probability has also been included for those who have little or no background knowledge in this area.

The main focus of the book is on practical management problems, but we have also considered theoretical issues where we feel that they are needed for readers to understand the scope and applicability of a particular technique. Many decision problems today are complicated by the need to consider a range of issues, such as those relating to the environment, and by the participation of divergent interest groups. To reflect this, we have included extensive coverage of problems involving multiple objectives and methods which are designed to assist groups of decision makers to tackle decision problems. An important feature of the book is the way in which it integrates the quantitative and psychological aspects of decision making. Rather than dealing solely with the manipulation of numbers, we have also attempted to address in detail the behavioral issues which are associated with the implementation of decision analysis. Throughout the book, reference is made to the latest computer software packages which can be used in the analysis of decision problems.

Besides being of interest to managers in general, the book is also intended for use as a main text on a wide range of courses. It is particularly suitable for people following courses in management and administration, such as an MBA, or final-year undergraduate programs in Business Studies, Quantitative Methods and Business Decision Analysis. Those studying for professional qualifications in areas like accountancy, where recent changes in syllabuses have placed greater emphasis on decision-making techniques, should also find the book useful. Almost all the chapters are followed by discussion questions or exercises, and we have included suggested answers to many of these exercises at the end of the book.

Paul Goodwin
George Wright

Acknowledgments

We would like to thank Larry Phillips for his advice, encouragement and the invaluable comments he made on a draft manuscript, and also Scott Barclay and Stephen Watson for their advice during the planning of this book.

Introduction

COMPLEX DECISIONS

Imagine that you are facing the following problem. For several years you have been employed as a manager by a major industrial company, but recently you have become dissatisfied with the job. You are still interested in the nature of the work and most of your colleagues have a high regard for you, but company politics are getting you down, and there appears to be little prospect of promotion within the foreseeable future. Moreover, the amount of work you are being asked to carry out seems to be increasing relentlessly and you often find that you have to work late in the evenings and at weekends.

One day you mention this to an old friend at a dinner party. 'There's an obvious solution,' he says. 'Why don't you set up on your own as a consultant? There must be hundreds of companies that could use your experience and skills, and they would pay well. I'm certain that you'd experience a significant increase in your income and there would be other advantages as well. You'd be your own boss, you could choose to work or take vacations at a time that suited you rather than the company and you'd gain an enormous amount of satisfaction from solving a variety of challenging problems.'

Initially, you reject the friend's advice as being out of the question, but as the days go by the idea seems to become more attractive. Over the years you have made a large number of contacts through your existing job and you feel reasonably confident that you could use these to build a client base. Moreover, in addition to your specialist knowledge and analytical ability you have a good feel for the way organizations tick, you are a good communicator and colleagues have often complimented you on your selling skills. Surely you would succeed.

However, when you mention all this to your spouse he or she expresses concern and points out the virtues of your current job. It pays well—enough for you to live in a large house in a pleasant neighborhood and to send the children to a good private school—and there are lots of other benefits such as health insurance and a company car. Above all, the job is secure. Setting up your own consultancy would be risky. Your contacts might indicate now that they could offer you plenty of work, but when it came to paying you good money would they really be interested? Even if you were to succeed eventually, it might take a while to build up a reputation, so would you be able to maintain your current lifestyle or would short-term sacrifices have to be made for long-term gains? Indeed, have you thought the idea through? Would you work from home or rent an office? After all, an office might give a more professional image to your business and increase your chances of success, but what would it cost? Would you employ secretarial staff or attempt to carry out this sort of work yourself? You are no typist and clerical work would leave less time for marketing your services and carrying out the consultancy itself. Of course, if you failed as a consultant, you might still get another job, but it is unlikely that it would be as well paid as your current post and the loss of self-esteem would be hard to take.

You are further discouraged by a colleague when you mention the idea during a coffee break. 'To be honest,' he says, 'I would think that you have less than a fifty–fifty chance of being successful. In our department I know of two people who have done what you're suggesting and given up after a year. If you're fed up here why don't you simply apply for a job elsewhere? In a new job you might even find time to do a bit of consultancy on the side, if that's what you want. Who knows? If you built up a big enough list of clients you might, in a few years' time, be in a position to become a full-time consultant, but I would certainly counsel you against doing it now.'

By now you are finding it difficult to think clearly about the decision; there seem to be so many different aspects to consider. You feel tempted to make a choice purely on emotional grounds—why not simply 'jump in' and take the risk?—but you realize that this would be unfair to your family. What you need is a method which will enable you to address the complexities of the problem so that you can approach the decision in a considered and dispassionate manner.

This is a personal decision problem, but it highlights many of the interrelated features of decision problems in general. Ideally, you would like to maximize your income, maximize your job security, maximize your job satisfaction, maximize your freedom and so on, so that the problem involves *multiple objectives*. Clearly, no course of action achieves all of these objectives, so you need to consider the trade-offs between the benefits offered by the various alternatives. For example, would the increased freedom of being your own boss be worth more to you than the possible short-term loss of income?

Second, the problem involves *uncertainty*. You are uncertain about the income that your consultancy business might generate, about the sort of work that you could get (would it be as satisfying as your friend suggests?), about the prospects you would face if the business failed and so on. Associated with this will be your *attitude to risk*. Are you a person who naturally prefers to select the least risky alternative in a decision or are you prepared to tolerate some level of risk?

Much of your frustration in attempting to understand your decision problem arises from its *complex structure*. This reflects, in part, the number of alternative courses of action from which you can choose (should you stay with your present job, change jobs, change jobs and become a part-time consultant, become a full-time consultant, etc.?), and the fact that some of the decisions are *sequential* in nature. For example, if you did decide to set up your own business should you then open an office and, if you open an office, should you employ a secretary? Equally important, have you considered all the possible options or is it possible to create new alternatives which may be more attractive than the ones you are currently considering? Perhaps your company might allow you to work for them on a part-time basis, allowing you to use your remaining time to develop your consultancy practice.

Finally, this problem is not yours alone; it also concerns your spouse, so the decision involves *multiple stakeholders*. Your spouse may view the problem in a very different way. For example, he or she may have an alternative set of objectives than you. Moreover, he or she may have different views of the chances that you will make a success of the business and be more or less willing than you to take a risk.

THE ROLE OF DECISION ANALYSIS

In the face of this complexity, how can decision analysis be of assistance? The key word is *analysis*, which refers to the process of breaking something down into its constituent parts. Decision analysis therefore involves the decomposition of a decision problem into a set of smaller (and, hopefully, easier to handle) problems. After each smaller problem has been dealt with separately, decision analysis provides a formal mechanism for integrating the results so that a course of action can be provisionally selected. This has been referred to as the 'divide and conquer orientation' of decision analysis.[1]

Because decision analysis requires the decision maker to be clear and explicit about his or her judgments it is possible to trace back through the analysis to discover why a particular course of action was preferred. This ability of decision analysis to provide an 'audit trail' means that it is possible to use the analysis to produce a defensible rationale for choosing a particular option. Clearly, this can be important when decisions have to be justified to senior staff, colleagues, outside agencies, the general public or even oneself.

When there are disagreements between a group of decision makers, decision analysis can lead to a greater understanding of each person's position so that there is a *raised consciousness* about the issues involved and about the root of any conflict. This enhanced communication and understanding can be particularly valuable when a group of specialists from different fields have to meet to make a decision. Sometimes the analysis can reveal that a disputed issue is not worth debating because a given course of action should still be chosen, whatever stance is taken in relation to that particular issue. Moreover, because decision analysis allows the different stakeholders to participate in the decision process and develop a shared perception of the problem it is more likely that there will be a *commitment* to the course of action which is eventually chosen.

The insights which are engendered by the decision analysis approach can lead to other benefits. Creative thinking may result so that new, and possibly superior, courses of action can be generated. The analysis can also provide guidance on what new information should be gathered before a decision is made. For example, is it worth undertaking more market research if this would cost $100 000? Should more extensive geological testing be carried out in a potential mineral field?

It should be stressed, however, that over the years the role of decision analysis has changed. No longer is it seen as a method for producing optimal solutions to decision problems. As Keeney[1] points out:

> Decision analysis will not solve a decision problem, nor is intended to. Its purpose is to produce insight and promote creativity to help decision makers make better decisions.

This changing perception of decision analysis is also emphasized by Phillips:[2]

> . . . decision theory has now evolved from a somewhat abstract mathematical discipline which when applied was used to help individual decision-makers arrive at optimal decisions, to a framework for thinking that enables different perspectives on a problem to be brought together with the result that new intuitions and higher-level perspectives are generated.

Indeed, in many applications decision analysis may be deliberately used to address only part of the problem. This *partial decision analysis* can concentrate on those elements of the problem where insight will be most valuable.

While we should not expect decision analysis to produce an optimal solution to a problem, the results of an analysis can be regarded as being 'conditionally prescriptive'. By this we mean that the analysis will show the decision maker what he or she should do, *given* the judgments which have been elicited from him or her during the course of the analysis. The basic assumption is that of *rationality*. If the decision maker is prepared to accept a set of rules (or axioms) which most people would regard as sensible then, to be rational, he or she should prefer the indicated course of action to its alternatives. Of course, the course of action prescribed by the analysis may well conflict with the decision maker's intuitive feelings. This conflict between the analysis and intuition can then be explored. Perhaps the judgments put forward by the decision maker represented only partially formed or inconsistent preferences, or perhaps the analysis failed to capture some aspect of the problem.

Alternatively, the analysis may enable the decision maker to develop a greater comprehension of the problem so that his or her preference changes towards that prescribed by the analysis. These attempts to explain why the rational option prescribed by the analysis differs from the decision maker's intuitive choice can therefore lead to the insight and understanding which, as we emphasized earlier, is the main motivation for carrying out decision analysis.

OVERVIEW OF THE BOOK

The book is organized in the following way. Chapter 2 shows how decision analysis can be used to handle problems which involve multiple objectives, but where there is little or no uncertainty about the outcomes of the different courses of action. Uncertainty is addressed in Chapter 3, where we show how probability theory can be used to measure uncertainty, and Chapter 4, where we apply probability to decision problems and illustrate a procedure for incorporating the decision maker's attitude to risk into the analysis.

As we saw at the start of this chapter, many decisions are difficult to handle because of their size and complex structure. In Chapters 5 and 6 we demonstrate methods which can help to clarify this complexity, namely decision trees, influence diagrams and simulation.

Of course, all decisions depend primarily on judgment: judgment about uncertainty, judgment about which alternative courses of action are available, judgment about the possible outcomes of a course of action and judgment about preferences. Decision analysis is not designed to replace these judgments but to provide a framework which will help decision makers to clarify and articulate them. In Chapter 7 we look at how a decision maker should revise judgments in the light of new information, while Chapters 8 and 9 consider psychological evidence on the quality of human judgment. The implications of this research are considered in Chapter 10, where we demonstrate techniques which have been developed to elicit probabilities from decision makers.

Although, in general, decisions made in organizations are ultimately the responsibility of an individual, often a group of people will participate in the decision-making process. Chapters 10 and 11 describe the role of decision analysis in this context with special emphasis on decision conferences and problems involving the allocation of resources.

Finally, in Chapter 13, we consider some alternative methods which have been developed to assist decision makers, such as expert systems, and we compare and contrast these with the decision analysis approach.

REFERENCES

1. Keeney, R. L. (1982) Decision Analysis: an Overview, *Operations Research*, **30**, 803–838.
2. Phillips, L. D. (1989) Decision analysis in the 1990's, in A. Shahini and R. Stainton (eds) *Tutorial Papers in Operational Research 1989*, Operational Research Society.

Chapter 2

Decisions Involving Multiple Objectives

INTRODUCTION

Many decision problems involve a number of objectives, and often these objectives conflict. For example, a local authority, faced with a decision on the route of a new road, might have to balance objectives such as minimizing cost and minimizing environmental damage. If the route incurring the lowest construction costs would also lead to the destruction of an important wildlife habitat then some judgment would have to be made about the relative importance of the objectives. Similarly, an individual choosing a house will have to balance factors such as cost, number of bedrooms, closeness to work, facilities available in the neighborhood and so on.

This chapter will examine how decisions involving multiple objectives can be analyzed. As we stated in Chapter 1, the central idea is that, by splitting the problem into small parts and focusing on each part separately, the decision maker is likely to acquire a better understanding of the problem than that which would be achieved by taking a holistic view. The unaided decision maker has 'limited information-processing capacity' (Wright[1]) and, when faced with a large and complex problem, there may be too much information for him to handle simultaneously. It can also be argued that, by requiring a commitment of time and effort, analysis encourages the decision maker to think deeply about his problem, enabling him to develop a rationale which is explicit and defensible. After such analysis the decision maker should be better able to explain and justify why a particular option is favored.

The methodology outlined in this chapter is underpinned by a set of axioms. We will discuss these towards the end of the chapter, but, for the moment, we can regard them as a set of generally accepted propositions or 'a formalization of common sense' (Keeney[2]). If the decision maker accepts the axioms then it follows that the results of the analysis will indicate how he should behave if he is rational. The analysis is therefore normative or prescriptive; it shows which alternative should be chosen if the decision maker acts consistently with his stated preferences.

The method explained here is normally applied in situations where a particular course of action is regarded as certain (or virtually certain) to lead to a given outcome so that uncertainty is not a major concern of the analysis. (We will consider, in later chapters, techniques which are designed to be used where risk and uncertainty are central concerns of the decision maker.) Nevertheless, there are exceptions to this rule, and we will show later how the method can be adapted to problems involving risk and uncertainty.

It should be emphasized that the main role of our analysis is to enable the decision maker to gain an increased understanding of his or her decision problem. If at the end of this analysis no single best course of action has been identified, this does not mean that the analysis was worthless. Often the insights gained may suggest other approaches to the problem or lead to a greater common understanding among a heterogeneous group of decision makers. They may lead to a complete reappraisal of the nature of the problem or enable a manager to reduce a large number of alternatives to a few, which can then be put forward to higher management with arguments for and against. Because of this, the analysis is flexible. Although we present it in this chapter as a series of stages, the decision maker is always free at any point to return to an earlier stage or to change the definition of the problem. Indeed, it is likely that this will happen as a deeper understanding of the nature of the problem is gained through the analysis.

BASIC TERMINOLOGY

Objectives and Attributes

Before proceeding, we need to clarify some of the basic terms we will be using. An *objective* has been defined by Keeney and Raiffa[3] as an indication of the preferred direction of movement. Thus, when stating objectives, we use terms like 'minimize' or 'maximize'. Typical objectives might be to minimize costs or maximize market share. An *attribute* is used to measure performance in relation to an objective. For example, if we have the objective 'maximize the exposure of a television advertisement' we may use the attribute 'number of people surveyed who recall seeing the advertisement' in order to measure the degree to which the objective was achieved. Sometimes we may have to use an attribute which is not directly related to the objective. Such an attribute is referred to as a *proxy attribute*. For example, a company may use the proxy attribute 'staff turnover' to measure how well they are achieving their objective of maximizing job satisfaction for their staff.

Value and Utility

For each course of action facing the decision maker we will be deriving a numerical score to measure its attractiveness to him. If the decision involves no element of risk and uncertainty we will refer to this score as the *value* of the course of action. Alternatively, where the decision involves risk and uncertainty, we will refer to this score as the *utility* of the course of action. Utility will be introduced in Chapter 4.

AN OFFICE LOCATION PROBLEM

The following problem will be used to illustrate the analysis of decisions involving multiple objectives. A small printing and photocopying business must move from its existing office because the site has been acquired for redevelopment. The owner of the business is considering seven possible new offices, all of which would be rented. Details of the location of these offices and the annual rent payable are given below.

Location of office	Annual rent ($)
Addison Square (A)	30 000
Bilton Village (B)	15 000
Carlisle Walk (C)	5 000
Denver Street (D)	12 000
Elton Street (E)	30 000
Filton Village (F)	15 000
Gorton Square (G)	10 000

While the owner would like to keep his costs as low as possible, he would also like to take other factors into account. For example, the Addison Square office is in a prestigious location close to potential customers, but it is expensive to rent. It is also an old, dark building which will not be comfortable for staff to work in. In contrast, the Bilton Village office is a new building which will provide excellent working conditions, but it is several miles from the center of town, where most potential customers are to be found. The owner is unsure how to set about making his choice, given the number of factors involved.

AN OVERVIEW OF THE ANALYSIS

The technique which we will use to analyse the office location problem is based on the Simple Multi-attribute Rating Technique (SMART), which was put forward by Edwards[4] in 1971. Because of the simplicity of both the responses required of the decision maker and the manner in which these responses are analyzed, SMART has been widely applied. The analysis involved is transparent, so the method is likely to yield an enhanced understanding of the problem and be acceptable to the decision maker who is distrustful of a mathematical 'black-box' approach. This, coupled with the relative speed by which the method can be applied, means that SMART has been found to be a useful vehicle for decision conferences (see Chapter 11), where groups of decision makers meet to consider a decision problem. The cost of this simplicity is that the method may not capture all the detail and complexities of the real problem. Nevertheless, in practice, the approach has been found to be extremely robust (see Watson and Buede[5]).

The main stages in the analysis are shown below:

Stage 1: *Identify the decision maker (or decision makers)*. In our problem we will assume that this is just the business owner, but in Chapter 12 we will look at the application of SMART to problems involving groups of decision makers.

Stage 2: *Identify the alternative courses of action*. In our problem these are, of course, represented by the different offices the owner can choose.

Stage 3: *Identify the attributes which are relevant to the decision problem*. The attributes which distinguish the different offices will be factors such as rent, size and quality of working conditions. In the next section we will show how a value tree can be of use when identifying relevant attributes.

Stage 4: For each attribute, assign values to *measure the performance of the alternatives on that attribute*. For example, how well do the offices compare when considering the quality of the working conditions they offer?

Stage 5: *Determine a weight for each attribute*. This may reflect how important the attribute is to the decision maker (though we will discuss the problem of using importance weights later).

Stage 6: *For each alternative, take a weighted average of the values assigned to that alternative*. This will give us a measure of how well an office performs over all the attributes.

Stage 7: *Make a provisional decision*.

Stage 8: *Perform sensitivity analysis* to see how robust the decision is to changes in the figures supplied by the decision maker.

CONSTRUCTING A VALUE TREE

Stages 1 and 2 of our analysis have already been completed: we know who the decision maker is and we have identified the courses of action open to him. The next step is to identify the attributes which the decision maker considers to be relevant to his problem. You will recall that an attribute is used to measure the performance of courses of action in relation to the objectives of the decision maker. This means that we need to arrive at a set of attributes which can be assessed on a numeric scale. However, the initial attributes elicited from the decision maker may be vague (e.g. he might say that he is looking for the office which will be 'the best for his business'), and they may therefore need to be broken down into more specific attributes before measurement can take place. A value tree can be useful here, and Figure 2.1 shows a value tree for this problem.

We start constructing the tree by addressing the attributes which represent the general concerns of the decision maker. Initially, the owner identifies two main attributes, which he decides to call 'costs' and 'benefits'. There is, of course, no restriction on the number of attributes which the decision maker can initially specify (e.g. our decision maker might have specified 'short-term costs', 'long-term costs', 'convenience of the move' and 'benefits' as his initial attributes). Nor is there any requirement to categorize the main attributes as costs and benefits. In some applications (e.g. Wooler and Barclay[6]) 'the risk of the options' is an initial attribute. Buede and Choisser[7] describe an engineering design application for the US Defense Communications Agency, where the main attributes are 'the effectiveness of the system' (i.e. factors such as quality of performance, survivability in the face of physical attack, etc.) and 'Implementation' (i.e. manning, ease of transition from the old system, etc.).

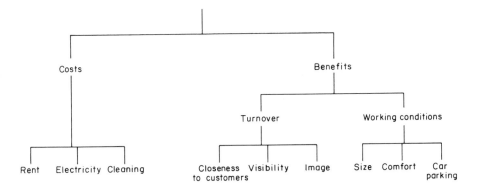

Figure 2.1 A value tree for the office location problem

Having established the main attributes for our business owner, we need to decompose them to a level where they can be assessed. The owner identifies three main costs that are of concern to him: the annual rent, the cost of electricity (for heating, lighting, operating equipment, etc.) and the cost of having the office regularly cleaned. Similarly, he decides that benefits can be subdivided into 'potential for improved turnover' and 'staff working conditions'. However, he thinks that he will have difficulty assessing each office's potential for improving turnover without identifying those attributes which will have an impact on turnover. He considers these attributes to be 'the closeness of the office to potential customers', 'the visibility of the site' (much business is generated from people who see the office while passing by) and 'the image of the location' (a decaying building in a back street may convey a poor image and lead to a loss of business). Similarly, the owner feels that he will be better able to compare the working conditions of the offices if he decomposes this attribute into 'size', 'comfort' and 'car parking facilities'.

Having constructed a value tree, how can we judge whether it is an accurate and useful representation of the decision maker's concerns? Keeney and Raiffa[3] have suggested five criteria which can be used to judge the tree:

(i) *Completeness.* If the tree is complete, all the attributes which are of concern to the decision maker will have been included.

(ii) *Operationality.* This criterion is met when all the lowest-level attributes in the tree are specific enough for the decision maker to evaluate and compare them for the different options. For example, if our decision maker felt that he was unable to judge the 'image' of the locations on a numeric scale, the tree would not be operational. In this case we could attempt to further decompose image into new attributes which were capable of being assessed, or we could attempt to find a proxy attribute for image.

(iii) *Decomposability.* This criterion requires that the performance of an option on one attribute can be judged independently of its performance on other attributes. In our problem, if the owner feels unable to assess the comfort afforded by an office without also considering its size, then decomposability has not been achieved, and we will need to look again at the tree to see if we can redefine or regroup these attributes.

(iv) *Absence of redundancy.* If two attributes duplicate each other because they actually represent the same thing then one of these attributes is clearly redundant. The danger of redundancy is that it may lead to double-counting, which may cause certain objectives to have undue weight when the final decision is made. One way of identifying redundancy is to establish whether the decision would in any way be affected if a given attribute was eliminated from the tree. If the deletion of the attribute would not make any difference to the choice of the best course of action then there is no point in including it.

(v) *Minimum size.* If the tree is too large any meaningful analysis may be impossible. To ensure that this does not happen, attributes should not be decomposed beyond the level where they can be evaluated. Sometimes the size of the tree can be reduced by eliminating attributes which do not distinguish between the options. For example, if all the offices in our problem offered identical car-parking facilities then this attribute could be removed from the tree.

Sometimes it may be necessary to find compromises between these criteria. For example, to make the tree operational it may be necessary to increase its size. Often several attempts at formulating a tree may be required before an acceptable structure is arrived at. This process of modification is well described in an application reported by Brownlow and Watson,[8] where a value tree was being used in a decision problem relating to the transportation of nuclear waste. The tree went through a number of stages of development as new insights were gained into the nature of the problem.

MEASURING HOW WELL THE OPTIONS PERFORM ON EACH ATTRIBUTE

Having identified the attributes which are of concern to the owner, the next step is to find out how well the different offices perform on each of the lowest-level attributes in the value tree. Determining the annual costs of operating the offices is relatively straightforward. The owner already knows the annual rent and he is able to obtain estimates of cleaning and electricity costs from companies which supply these services. Details of all these costs are given in Table 2.1.

Table 2.1 Costs associated with the seven offices

Office	Annual rent ($)	Annual cleaning costs ($)	Annual electricity costs ($)	Total cost ($)
Addison Square	30 000	3 000	2 000	35 000
Bilton Village	15 000	2 000	800	17 800
Carlisle Walk	5 000	1 000	700	6 700
Denver Street	12 000	1 000	1 100	14 100
Elton Street	30 000	2 500	2 300	34 800
Filton Village	15 000	1 000	2 600	18 600
Gorton Square	10 000	1 100	900	12 000

At a later stage in our analysis we will need to trade off the costs against the benefits. This can be an extremely difficult judgment to make. Edwards and Newman[9] consider this kind of judgment to be 'the least secure and most uncomfortable to make' of all the judgments required in decisions involving multiple objectives. Because of this we will now ignore the costs until the end of our analysis and, for the moment, simply concentrate on the benefit attributes.

In measuring these attributes our task will be made easier if we can identify variables to represent the attributes. For example, the size of an office can be represented by its floor area in square feet. Similarly, the distance of the office from the town center may provide a suitable approximation for the attribute 'distance from potential customers'. However, for other attributes such as 'image' and 'comfort' it will be more difficult to find a variable which can be quantified. Because of this, there are two alternative approaches which can be used to measure the performance of the offices on each attribute: direct rating and the use of value functions.

Direct Rating

Let us first consider those attributes which cannot be represented by easily quantifiable variables, starting with the attribute 'image'. The owner is first asked to rank the locations in terms of their image from the most preferred to the least preferred. His rankings are:

1. Addison Square
2. Elton Street
3. Filton Village
4. Denver Street
5. Gorton Square
6. Bilton Village
7. Carlisle Walk.

Addison Square, the best location for image, can now be given a value for image of 100 and Carlisle Walk, the location with the least appealing image, can be given a value of 0. As we explain below, any two numbers could have been used here as long as the number allocated to the most-preferred location is higher than that allocated to the least preferred. However, the use of 0 and 100 makes the judgments which follow much easier and it also simplifies the arithmetic.

The owner is now asked to rate the other locations in such a way that the space between the values he gives to the offices represents his strength of preference for one office over another in terms of image. Figure 2.2. shows the values allocated by the owner. This shows that the *improvement* in image

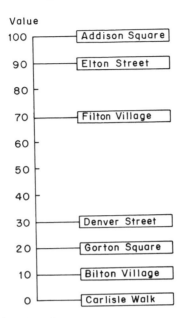

Figure 2.2 A value scale for office image

between Carlisle Walk and Gorton Square is perceived by the owner to be twice as preferable as the improvement in image between Carlisle Walk and Bilton Village. Similarly, the improvement in image between Carlisle Walk and Addison Square is seen to be ten times more preferable than the improvement between Carlisle Walk and Bilton Village.

Note that it is the *interval* (or improvement) between the points in the scale which we compare. We cannot say that the image of Gorton Square is twice as preferable as that of the Bilton Village office. This is because the allocation of a zero to represent the image of Carlisle Walk was arbitrary, and we therefore have what is known as an *interval scale*, which allows only intervals between points to be compared. The Fahrenheit and Celsius temperature scales are the most well-known examples of interval scales. We cannot, for example, say that water at 80 °C is twice the temperature of water at 40 °C. You can verify this by converting the temperatures to degrees Fahrenheit to obtain 175 °F and 104 °F, respectively. Clearly, the first temperature is no longer twice the second temperature. However, we can say that an increase in temperature from 40° to 80 °C is twice that of an increase from 40° to 60 °C. You will find that such a comparison does apply, even if we convert the temperatures to degrees Fahrenheit.

Having established an initial set of values for image, these should be checked to see if they consistently represent the preferences of the decision maker. We can achieve this by asking him, for example, if he is happy that the improvement in image between Elton Street and Addison Square is roughly as preferable as the improvement in image between Gorton Square and Denver Street. Similarly, is he happy that the improvement in image between Carlisle Walk and Denver Street is less preferable than that between Denver Street and Elton Street? The answers to these questions may lead to a revision of the values. Of course, if the owner finds it very difficult to make these sorts of judgments we may need to return to the value tree and see if we can break image down into more measurable attributes. Nevertheless, it should be emphasized that the numbers allocated by the owner to the different offices do not need to be precise. As we will see later, the choice of a course of action is generally fairly robust, and it often requires quite substantial changes in the figures supplied by the decision maker before another option is preferred.

This procedure for obtaining values can be repeated for the other less easily quantified attributes. The values allocated by the owner for the attributes 'comfort', 'visibility' and 'car-parking facilities' are shown in Table 2.2 (see page 22).

Value Functions

Let us now consider the benefit attributes which can be represented by easily quantified variables. First, we need to measure the owner's relative strength of preference for offices of different sizes. The floor area of the offices is shown below.

		Floor area (ft^2)
Addison Square	(A)	1000
Bilton Village	(B)	550
Carlisle Walk	(C)	400
Denver Street	(D)	800
Elton Street	(E)	1500
Filton Village	(F)	400
Gorton Square	(G)	700

Now it may be that an increase in area from 500 ft^2 to 1000 ft^2 is very attractive to the owner, because this would considerably improve working conditions. However, the improvements to be gained from an increase from 1000 ft^2 to 1500 ft^2 might be marginal and make this increase less attractive. Because of this, we need to translate the floor areas into values. This can be achieved as follows.

The owner judges that the larger the office, the more attractive it is. The largest office, Elton Street, has an area of 1500 ft^2 so we can give 1500 ft^2 a value of 100. In mathematical notation we can say that:

$$v(1500) = 100$$

where $v(1500)$ means 'the value of 1500 ft^2'. Similarly, the smallest offices (Carlisle Walk and Filton Village) both have areas of 400 ft^2 so we can attach a value of 0 to this area, i.e. $v(400) = 0$.

We now need to find the value of the office areas which fall between the most-preferred and least-preferred areas. We could ask the owner to directly rate the areas of the offices under consideration using the methods of the previous section. However, because areas involving rather awkward numbers are involved, it may be easier to derive a value function. This will enable us to estimate the values of any office area between the most and least preferred. There are several methods which can be used to elicit a value function, but one of the most widely applied is *bisection*.

This method requires the owner to identify an office area whose value is halfway between the least-preferred area (400 ft^2) and the most-preferred area (1500 ft^2). Note that this area does not necessarily have to correspond to that of one of the offices under consideration. We are simply trying to elicit the owner's preferences for office areas in general, and having obtained this information we can then use it to assess his preference for the specific office areas which are available to him. Initially, the owner suggests that the midpoint area would be 1000 ft^2. This implies that an increase in area from 400 ft^2 to 1000 ft^2 is just as attractive as an increase from 1000 ft^2 to 1500 ft^2. However, after some thought he rejects this value. The increases from smaller areas will, he reasons, reduce over-crowding and so be much more attractive than increases from larger areas which would only lead to minor improvements. He is then offered other candidates for the midpoint position (for example, 900 ft^2 and 600 ft^2), but rejects these values as well. Finally, he agrees that 700 ft^2 has the midpoint value, so $v(700) = 50$.

Having identified the midpoint value, the decision maker is now asked to identify the 'quarter points'. The first of these will be the office area, which has a value halfway between the least-preferred area (400 ft^2) and the midpoint area (700 ft^2). He decides that this is 500 ft^2, so $v(500) = 25$. Similarly, we ask him to identify an area which has a value halfway between the midpoint area (700 ft^2) and the best area (1500 ft^2). He judges this to be 1000 ft^2, which implies that $v(1000) = 75$. We now have the values for five floor areas and this enables us to plot the value function for office size, which is shown in Figure 2.3. This value function can be used to estimate the values for the actual areas of the offices under consideration. For example, the Bilton Village office has an area of 550 ft^2 and the curve suggests that the value of this area is about 30.

A similar method can be applied to the attribute 'closeness to customers'. This attribute has been represented by the variable 'distance from town center' and the value function is shown in Figure 2.4. Note that the greater the distance from the town center, the lower the value will be. The curve also suggests that a move from 0 to 2 miles from the town center is far more damaging to business than a move from 6 to 8 miles. The values identified for the seven offices in terms of 'office area' and 'closeness to customers' are shown in Table 2.2 (see page 22).

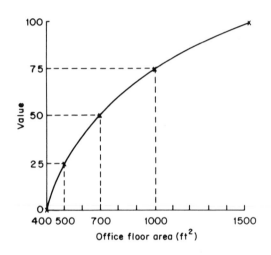

Figure 2.3 Constructing a value function for office floor area

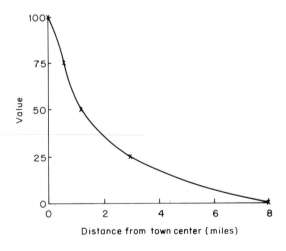

Figure 2.4 A value function for distance from customers

DETERMINING THE WEIGHTS OF THE ATTRIBUTES

In the previous section we derived values to measure how well each office performed on each attribute. For example, we found that Addison Square was the best office for image and closeness to customers, but it was the least-preferred office for providing comfortable working conditions for staff. Clearly, in order to make a decision the owner now needs to combine the values for the different attributes in order to gain a view of the overall benefits which each office has to offer.

An intuitively appealing way of achieving this is to attach weights to each of the attributes which reflect their importance to the decision maker. For example, he might consider office area to be less important than distance from customers and therefore give a weight of only 1 to office area and a weight of 5 to distance from customers. The problem with this approach is that it may not take into account how large the range is between the most-preferred and least-preferred options on each attribute. To illustrate this, consider the following problem.

A civil engineering company is anxious to complete a major project for an important client as quickly as possible. By spending more on the project (e.g. by hiring extra equipment and labor) time can be saved and the company has to decide how much extra to spend. This extra expenditure can range from $0 to $25 million and the resulting time saved can range from 0 to 200 days. The project leader considers that 'days saved' is four times more important than the extra expenditure this would require. Suppose that subsequently it is found that, for technical reasons, the number of days saved can only range from 0 to 2, though $25 million would still be required to make the maximum 2-day saving. Almost certainly, the project leaders should now change his weights, otherwise he is implying that a mere 2-day saving is still four times more important than expenditure of $25 million. Thus, if importance weights are used, they should be adjusted so that the smaller the range over which the attribute is assessed, the smaller the importance weight which should be used. However, there is no clear evidence from research that people do take the range into account when assigning importance weights (see von Winterfeldt and Edwards[10]).

Fortunately, this problem can be avoided by using *swing weights*. These are derived by asking the decision maker to compare a change (or swing) from the least-preferred to the most-preferred value on one attribute to a similar change in another attribute. The simplest approach is to proceed as follows. Consider the lowest-level attributes on the 'Benefits' branch of the value tree (Figure 2.1). The owner is asked to imagine a hypothetical office with all these attributes at their least-preferred levels, that is, an office which is the greatest distance (i.e. 8 miles) from the town center, has the worst position for visibility and has the worst image, the smallest size and so on.

Then he is asked, if just one of these attributes could be moved to its best level, which would he choose? The owner selects 'closeness to customers'. After this change has been made, he is asked which attribute he would next choose to move to its best level, and so on until all the attributes have been ranked. The owner's rankings are:

1. Closeness to customers
2. Visibility
3. Image
4. Size
5. Comfort
6. Car-parking facilities

We can now give 'closeness to customers' a weight of 100. The other weights are assessed as follows. The owner is asked to compare a swing from the least visible location to the most visible, with a swing from the most distant location from customers to the closest location. After some thought, he decides that the swing in 'visibility' is 80% as important as the swing in 'closeness to customers' so visibility is given a weight of 80. Similarly, a swing from the worst 'image' to the best is considered to be 70% as important as a swing from the worst to the best location for 'closeness to customers', so 'image' is assigned a weight of 70. The procedure is repeated for all the other lower-level attributes and Figure 2.5 illustrates the results. As shown below, the six weights obtained sum to 310, and it is conventional to 'normalize' them so that they add up to 100 (this will make

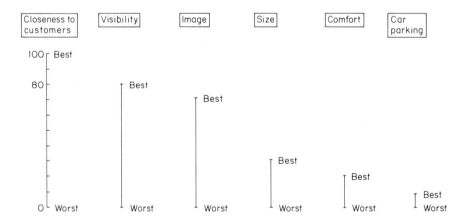

Figure 2.5 Derivation of swing weights. For example, a swing from the worst to the best location for visibility is considered to be 80% as important as a swing from the worst to the best location for closeness to customers.

later stages of the analysis easier to understand). Normalization is achieved by simply dividing each weight by the sum of the weights (310) and multiplying by 100.

Attribute	Original weights	Normalized weights (to nearest whole number)
Closeness to customers	100	32
Visibility	80	26
Image	70	23
Size	30	10
Comfort	20	6
Car-parking facilities	10	3
	310	100

The weights for the higher-level attributes in the value tree, 'turnover' and 'working conditions', are now found by summing the appropriate lower-level weights, so the weight for turnover is 81 (i.e. 32+26+23) and the weight for working conditions is 19 (i.e. 10+6+3).

AGGREGATING THE BENEFITS USING THE ADDITIVE MODEL

We now have (1) a measure of how well each office performs on each attribute and (2) weights which enable us to compare the values allocated to one attribute with the values allocated to the others. This means that we are now in a position to find out how well each office performs overall by combining the six value scores allocated to that office.

To do this, we will assume that the additive model is appropriate. As we show below, this simply involves adding an office's weighted value scores together to obtain a measure of the overall benefits which that office has to offer. The additive model is by far the most widely used, but it is not suitable for all circumstances. In particular, the model is inappropriate where there is an interaction between the values associated with some of the attributes. For example, when choosing a house, an attractive architecture and a pleasant garden may complement each other, leading to a combined value which is greater than the sum of the individual values. We will examine the limitations of the additive model later.

The calculations which the additive model involves are shown below for Addison Square. Each value is multiplied by the weight attached to that attribute. The resulting products are then summed and divided by 100 to obtain the overall value of benefits at that location.

Attribute	Addison Square values	Weight	Value × weight
Closeness to customers	100	32	3200
Visibility	60	26	1560
Image	100	23	2300
Size	75	10	750
Comfort	0	6	0
Car-parking facilities	90	3	270
			8080

Therefore the aggregate value for Addison Square is 8080/100, i.e. 80.8. Table 2.2 gives a summary of the values obtained for all the offices and their aggregate values. It can be seen that Addison Square has the highest value for benefits and Filton Village the lowest. However, so far we have ignored the costs associated with the offices and the next section shows how these can be taken into account.

Table 2.2 Values and weights for the office location problem

Attribute	Weight	Office A	B	C	D	E	F	G
Closeness	32	100	20	80	70	40	0	60
Visibility	26	60	80	70	50	60	0	100
Image	23	100	10	0	30	90	70	20
Size	10	75	30	0	55	100	0	50
Comfort	6	0	100	10	30	60	80	50
Car parking	3	90	30	100	90	70	0	80
Aggregate Benefits		80.8	39.4	47.4	52.3	64.8	20.9	60.2

TRADING BENEFITS AGAINST COSTS

You will recall that until now we have ignored the costs of the offices because of the difficulties which decision makers often have in making judgments about the trade-off between costs and benefits. If our decision maker had not found this to be a problem then we could have treated cost as just another attribute. We could therefore have allocated values to the various costs, with a value of 100 being given to the office which had the lowest costs and 0 to the office with the highest. Weights could then have been derived to compare swings from the least-preferred to the most-preferred level of benefits with swings from the worst to the best cost. This would have enabled the value for cost to have been included when we used the additive model to derive an overall value to measure the attractiveness of the different offices. The office achieving the highest overall value would have been the one the owner should choose. This approach has been used in problems where decision makers experienced no difficulty in assigning weights to all the attributes.

However, because our owner does have difficulties in judging the cost–benefit trade-off, we can proceed as follows. In Figure 2.6 the aggregate value of benefits has been plotted against the annual cost for each of the offices. Note that the cost scale has been 'turned round', so that the lower (and therefore more preferable) costs are to the right. This is to make this graph comparable with ones we will meet later in the book. Clearly, the higher an office appears on the benefits scale and the further to the right on the cost scale, the more attractive it will be. If we compare Addison Square (A) with Elton Street (E) it can be seen that, while both have similar costs, Addison Square has higher benefits. It would not therefore be worth considering Elton Street, and this office is said to be *dominated* by Addison Square. Similarly Gorton Square (G) not only has lower costs but also higher benefits compared with Bilton Village (B), Denver Street (D) and Filton Village (F). Therefore B, D, and F are also dominated offices. Thus the only locations which are worth considering are Addison Square (A), Gorton Square (G) and Carlisle Walk (C). These non-dominated offices are said to lie on the *efficient frontier*.

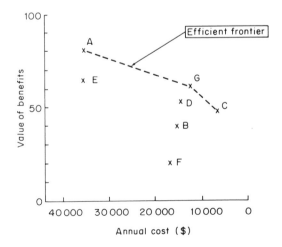

Figure 2.6 A plot of benefits against costs for the seven offices

The choice between the three offices on the efficient frontier will depend on the relative weight the owner attaches to costs and benefits. If he is much more concerned about benefits, then Addison Square will be his choice. Alternatively, if he is more concerned to keep his costs low, then he should choose Carlisle Walk. Gorton Square would be an intermediate choice. It costs $5300 more per year than Carlisle Walk, but offers slightly higher benefits.

This information may be sufficient for the owner to make a choice. At the very least, it should illuminate his understanding of the decision problem. He may be surprised that Bilton Village has fared so badly or that Carlisle Walk has done so well, and he may wish to check back through the data he has supplied to see why this has happened.

However, it is possible that the decision maker still feels unable to choose between the three offices on the efficient frontier and thinks that a more formal approach would help him. If this is the case then the following procedure suggested by Edwards and Newman[9] can be used.

Consider first a move from Carlisle Walk (C) to Gorton Square (G). This would lead to an increase in the value of benefits from 47.4 to 60.2, an increase of 12.8. However, it would also lead to an increase in costs of $5300. Therefore each one-point increase in the value of benefits would cost him $5300/12.8, which is $414. Similarly, a move from Gorton Square (G) to Addison Square (A) would increase the value of benefits by 20.6 points at an extra cost of $23 000. This would therefore cost $23 000/20.6, which is $1117 for each extra benefit value point. So if an extra value point is worth less than $414 to the owner he should choose Carlisle Walk. If it is worth between $414 and $1117, he should choose Gorton Square, and if it is worth paying more than $1117 for each extra value point he should choose Addison Square.

Now we need to determine how much each extra value point is worth to the owner. This can be achieved by selecting a lower-level attribute from the value tree which the owner will find fairly easy to evaluate in monetary terms. The owner suggests that this is 'image'. He is then asked what extra annual cost he would be prepared to incur for a move from the office with the worst image to one with the best. He answers that it would be worth paying an extra $15 000. This means that he considers that it would be worth paying $15 000 for a 100-point increase in the value of image. Now the weight of image is 23% of the total weight allocated to the attributes. So an increase of 100 points on the image scale would increase the aggregate value of benefits by 23 points. Therefore the owner is prepared to pay $15 000 to gain 23 points in the value of aggregate benefits. This implies that he is prepared to pay $15 000/23 or $652 per point. On this basis he should choose the Gorton Square office.

Of course, the data we have been dealing with are far less precise than the above analysis might have implied, and it is unlikely that the owner will be 100% confident about the figures which he has put forward. Before making a firm recommendation therefore we should explore the effect of changes in these figures. This topic is covered in the next section.

SENSITIVITY ANALYSIS

Sensitivity analysis is used to examine how robust the choice of an alternative is to changes in the figures used in the analysis. The owner is a little worried about the weight of turnover (i.e. 81) relative to working conditions (i.e. 19) and he would like to know what would happen if this weight was changed. Figure 2.7 shows how the value of benefits for the different offices varies with changes in the weight placed on turnover. For example, if turnover had a weight of zero this would imply that the three turnover attributes would also have zero weights, so the weights for the six lowest-level benefit attributes would now be: closeness to customers 0, visibility 0, image 0, size 30, comfort 20, car parking 10. These normalize to 0, 0, 0, 50, 33.3, and 16.7, respectively, which would mean that Elton Street (E), for example, would have benefits with a value of 81.7. At the other extreme, if turnover had a weight of 100 (and therefore working conditions a weight of zero) the value of benefits for Elton Street would have been 60.4. The line joining these points shows the value of benefits for Elton Street for turnover weights between 0 and 100.

It can be seen that Elton Street gives the highest value of benefits as long as the weight placed on turnover is less than 52.1. If the weight is above this figure then Addison Square (A) has the highest value of benefits. Since the owner assigned a weight of 81 to turnover, it will take a fairly large change in this weight before Elton Street is worth considering, and the owner can be reasonably confident that Addison Square should appear on the efficient frontier.

Figure 2.7 also shows that no change in the weight attached to turnover will make the other offices achieve the highest value for benefits. Filton Village (F), in particular, scores badly on any weight. If we consider the other two offices on the efficient frontier we see that Gorton Square (G) always has higher-valued benefits than Carlisle Walk (C).

Similar analysis could be carried out on the lower-level weights. For example, the owner may wish to explore the effect of varying the weights attached to 'closeness to customers' and 'visibility' while keeping the weight attached to 'image' constant. Carrying out sensitivity analysis should contribute to the decision maker's understanding of his problem and it may lead him to reconsider some of the figures he has supplied. In many cases sensitivity analysis also shows that the data supplied do not need to be precise. As we saw above, large changes in these figures are often required

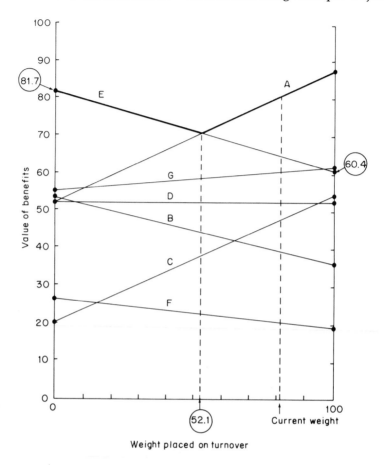

Figure 2.7 Sensitivity analysis for weight placed on turnover

before one option becomes more attractive than another: a phenomenon referred to as 'flat maxima' by von Winterfeldt and Edwards.[10]

THEORETICAL CONSIDERATIONS

The Axioms of the Method

In our analysis of the office location problem we implicitly made a number of assumptions about the decision maker's preferences. These assumptions, which are listed below, can be regarded as the axioms of the method. They represent a set of postulates which may be regarded as reasonable. If the decision maker accepts these axioms, and if he is rational (i.e. if he behaves consistently in relation to the axioms), then he should also accept the preference rankings indicated by the method. Let us now consider the axioms:

(1) *Decidability*: We assumed that the owner was able to decide which of two options he preferred. For example, we assumed that he could state whether the improvement in image between Carlisle Walk and Gorton Square was greater than the improvement between Carlisle Walk and Bilton Village. It may have been that the owner was very unsure about making this comparison or he may have refused to make it at all.

(2) *Transitivity*: The owner preferred the image of Addison Square to Bilton Village (i.e. A to B). He also preferred the image of Bilton Village to Carlisle Walk (i.e. B to C). If transitivity applies then the owner must therefore also prefer the image of Addison Square to Carlisle Walk (i.e. A to C).

(3) *Summation*: This implies that if the owner prefers A to B and B to C, then the strength of preference of A over C must be greater than the strength of preference of A over B (or B over C).

(4) *Solvability*: This assumption was necessary for the bisection method of obtaining a value function. Here the owner was asked to identify a distance from the center of town which had a value halfway between the worst and best distances. It was implicitly assumed that such a distance existed. In some circumstances there may be 'gaps' in the values which an attribute can assume. For example, the existence of a zone of planning restrictions between the center of the town and certain possible locations might mean that siting an office at a distance which has a value halfway between the worst and best distances is not a possibility which the decision maker can envisage.

(5) *Finite upper and lower bounds for value*: In assessing values we had to assume that the best option was not so wonderful and the worst option was not so awful that values of plus and minus infinity would be assigned to these options.

Assumptions Made when Aggregating Values

In our analysis we used the additive model to aggregate the values for the different attributes. As we pointed out, the use of this model is not appropriate where there is an interaction between the scores on the attributes. In technical terms, in order to apply the model we need to assume that *mutual preference independence* exists between the attributes.

To demonstrate preference independence let us suppose that our office location problem only involves two attributes: 'distance from customers' and 'office size'. Our decision maker is now offered two offices, X and Y.

These are both the same size (1000 ft²) but X is closer to customers, as shown below:

Office	Distance from customers	Office floor area
X	3 miles	1000 ft²
Y	5 miles	1000 ft²

Not surprisingly, the decision maker prefers X to Y. Now suppose that we change the size of both offices to 400 ft². If, as is likely, the decision maker still prefers X to Y his preference for a distance of 3 miles over a distance of 5 miles has clearly been unaffected by the change in office size. This might remain true if we change the size of both offices to any other possible floor area. If this is the case, we can say that 'distance from customers' is preference independent of 'office size' because the preference for one distance over another does not depend on the size of the offices.

If we also found that 'size of office' is preference independent of 'distance from customers' then we can say that the two attributes are *mutually preference independent*. Note that mutual preference independence does not automatically follow. When choosing a holiday destination, you may prefer a warmer climate to a cooler one, irrespective of whether or not the hotel has an open-air or indoor swimming pool. However, your preference between hotels with open-air or indoor swimming pools will probably depend on whether the local climate is warm or cool.

To see what can happen when the additive model is applied to a problem where mutual preference independence does not exist, consider the following problem. Suppose now that our office location decision depends only on the attributes 'image' and 'visibility' and the owner has allocated weights of 40 and 60 to these two attributes. Two new offices, P and Q, are being compared and the values assigned to the offices for each of these attributes are shown below (0=worst, 100=best).

Office	Visibility	Image
P	0	100
Q	100	0

Using the additive model, the aggregate value of benefits for P will be:

$$(40 \times 0) + (60 \times 100) = 6000 \text{ i.e. } 60 \text{ after dividing by } 100$$

while the aggregate value of benefits for Q will be:

$$(40 \times 100) + (60 \times 0) = 4000 \text{ i.e. } 40 \text{ after dividing by } 100$$

This clearly suggests that the decision maker should choose office P. However, it may be that he considers image only to be of value if the office is highly visible. Office P's good image is, he thinks, virtually worthless because it is not on a highly visible location and he might therefore prefer office Q. Thus, if image is not preference independent of visibility, the additive model will not correctly represent the owner's preferences.

How can the absence of mutual preference independence be identified? The most obvious way in which this will reveal itself is in the use of phrases like 'this depends on. . . ' when the decision maker responds to questions. For example, when asked to assign a value to the 'image' of an office, our decision maker might well have said 'that depends on how visible the office is'.

If mutual preference independence does not exist it is usually possible to return to the value tree and redefine the attributes so that a set of attributes which are mutually preference independent can be identified. For example, perhaps visibility and image could be replaced by a single attribute 'ability to attract casual customers'.

In the occasional problems where this is not possible, other models are available which can handle the interaction between the attributes. The most well known of these is the multiplicative model. Consider again the case of the house purchase decision where the quality of the architecture and attractiveness of the garden complemented each other. If we let $V(A)$ = the value of the architecture of a given house and $V(G)$ = a value for the attractiveness of the garden then we might find that the following represented the overall value of the house:

$$\text{Value} = 0.6V(A) + 0.3V(G) + 0.1V(A)V(G)$$

The numbers in the above expression represent the weights (note that they sum to 1) and the last expression, which involves multiplying the values together, represents the interaction between architecture and garden. Because the multiplicative model is not widely used we will not consider it in detail. Longer discussions can be found in Bodily[11] and von Winterfeldt and Edwards.[10]

CONFLICTS BETWEEN INTUITIVE AND ANALYTIC RESULTS

It may be that, if the decision maker had viewed the problem holistically, then he would have ranked his preferences for the offices in a very different order from that obtained through our analysis. This could be because the problem was too large and complex for him to handle as a whole so that his true perferences were not reflected in his holistic judgments. An analogy can be made with an attempt to answer a mathematical problem by using mental arithmetic rather than a calculator. This view is supported by research which suggests that the correlation of preference rankings derived from holistic judgments with those derived from SMART-type analyses decreases as the number of attributes in the problem gets larger. In other words, the larger the problem, the less reliable holistic judgments may be (see von Winterfeldt and Edwards[10] for a summary of this research).

Alternatively, discrepancies between holistic and analytic results may result when the axioms are not acceptable to the decision maker. It is possible that the decision maker could argue the case for a different set of sensible axioms. As long as he behaved consistently with these axioms, we could not argue that his rejection of the results of our analysis was irrational.

Nevertheless, any conflict between holistic and analytic rankings should be examined, since it may suggest that an important element of the problem has not been captured by the analysis. For example, an important attribute may have been left out or the interaction between two attributes may not have been taken into account. We can, of course, never be certain that a decision model is a faithful representation of the decision maker's preferences. In a computer model of a traffic system, for example, the model's validity can be assessed by comparing its predictions with the behavior of the real system, but here we are attempting to model the decision maker's beliefs and attitudes for which there is no physical analog. This begs the question: at what point do we decide that a decision model is adequate so that further refinements and revisions are not worth carrying out?

One approach to this question is to use Phillip's[12] concept of a *requisite decision model*. We will examine this idea in more detail in Chapter 11 in the context of group decision making, but briefly, a model is considered to be requisite when it provides the decision maker with enough guidance and insight to decide upon a course of action. Thus at the point where the decision maker knows what to do next a requisite model has been achieved. Phillips argues that:

> the modeling process uses the sense of unease among the problem owners about the results of the current model as a signal that further modeling may be needed, or that intuition may be wrong. If exploration of the discrepancy between holistic judgment and model results shows the model to be at fault, then the model is not requisite—it is not yet sufficient to solve the problem. The model can be considered requisite only when no new intuitions emerge about the problem.

Thus the requisite modeling process does not attempt to obtain an exact representation of the decision maker's beliefs and preferences, or to prescribe an optimal solution to his problem. However, by exploiting the conflicts between the results of the analysis and his intuitive judgments it will help him to resolve conflicts and inconsistencies in his thinking. As a deeper understanding of the problem is obtained the model will be revised and the discrepancy between the analytical and intuitive judgments will be reduced. Eventually, the decision maker will find that the model provides enough guidance for him to reach a decision[13].

SUMMARY

In this chapter we have looked at a method for analyzing decision problems where each alternative had several attributes associated with it. This meant that the performance of each alternative had to be measured on each attribute and then the attributes themselves had to be 'weighed against' each other before a decision could be made. The central idea was that, by splitting the problem into small parts and focusing on each part separately, the decision maker was likely to acquire a better understanding of his problem than he would have achieved by taking a holistic view. We saw that the method required the decision maker to quantify his strengths of preferences. While this may not have been an easy process, we found that the figures put forward did not need to be exact, though we did try to ensure that they were consistent.

The decision problem presented in this chapter was designed to be amenable to hand calculations. This would, however, be an extremely tedious way of approaching larger problems, and for these it would be necessary to use a computer. One package that is available for this purpose is HIVIEW, which was developed by the London School of Economics' Decision Analysis Unit (see Barclay[13]). This has all the features which are needed to carry out SMART-type analysis, including a facility which allows the user to construct and modify value trees on-screen.

We stated in the Introduction that this method is normally applied where risk and uncertainty are not major concerns of the decision maker. However, it is possible to apply the method even in these circumstances by treating risk as an attribute. Wooler and Barclay[6] describe such an application involving a strike-prone production facility. (The analysis involved a group of managers in a decision conference.) A value tree was used to decompose 'risk' into lower-level attributes such as 'risk of strikes' and 'public relations risks', and the various strategies were scored in terms of their performance on these attributes using direct rating (for example, the least risky option was allocated the highest value). A second part of the value tree dealt with the benefits of the strategies and these were similarly scored. A graph such as Figure 2.7 was then used to display the aggregate risk of strategies against their aggregate benefits (rather than costs against benefits). We will consider a similar approach to risk in the context of a group decision problem in

Chapter 12. However, a number of techniques have been specially designed to handle decisions involving a large element of risk and uncertainty, and we will consider these methods in the following chapters.

EXERCISES

(1) Formulate a value tree to identify the attributes which are of concern to you when choosing a vacation.

(2) You need a word-processing package for the personal computer in your office. Because your employer will pay for the package you are not concerned about the cost, but you would like a package which is as easy to use as possible and which also has a wide range functions such as a thesaurus, spell checker and graphics. After discussing the matter with a friend who is something of an expert in this field, you identify seven potential packages and allocate values to them to reflect their ease of use and available facilities. These values are shown below (0=worst, 100=best).

Package	Ease of use	Facilities available
Super Quill	100	30
Easywrite	90	70
Wordright	50	20
Lexico	0	40
Ultraword	20	100
Keywrite	40	0
Fastwrite	85	55

(a) Plot each package's value for 'ease of use' and 'facilities available' on a graph and hence determine the packages which lie on the efficient frontier.

(b) Suppose that you judge that a switch from a package with the least facilities available to one with the most facilities is only 60% as attractive as a switch from a package which is least easy to use to one which is the most easy to use. Assuming that mutual preference independence exists between the two attributes, which package should you choose?

(c) After some reflection you realize that the extra facilities available on a package will be of little value to you if they are going to be difficult to use. What does this imply about your method of analysis in (b)?

(3) A chemical company is expanding its operations and a disused woollen mill is to be converted into a processing plant. Four companies have submitted designs for the equipment which will be installed in the mill and a choice has to be made between them. The manager of the chemical company has identified three attributes which he considers to be important in the decision: 'cost', 'environmental impact' and 'reliability'. He has

assessed how well each design performs on each attribute by allocating values on a scale from 0 (the worst design) to 100 (the best). These values are shown below, together with the costs which will be incurred if a design is chosen.

		Benefits	
		Environmental	
Design	Costs ($)	impact	Reliability
A	90 000	20	100
B	110 000	70	0
C	170 000	100	90
D	60 000	0	50

(a) The manager is having difficulty in allocating weights to the two benefit attributes. Assuming that the two weights sum to 100 and that mutual preference independence exists between the attributes, perform a sensitivity analysis to show how the design offering the highest value for aggregate benefits will vary depending upon the weight which has been allocated to 'environmental impact'.

(b) Eventually, the manager decides to allocate 'environmental impact' a weight of 30 and 'reliability' a weight of 70. By plotting the benefits and costs of the designs on a graph, identify the designs which lie on the efficient frontier.

(c) The manager also decides that if he was offered a hypothetical design which had the lowest reliability and the worst environmental impact he would be prepared to pay $120 000 to convert that design to one which had the best impact on the environment but which still had the lowest level of reliability. Which design should the manager choose?

(4) A British company has won an important contract to supply components regularly to Poland. Four methods of transport are being considered: (i) air, (ii) sea, (iii) road and ferry and (iv) rail and ferry. The company's distribution manager has identified four relevant attributes for the decision: Punctuality, Safety of Cargo, Convenience and Costs. She has also allocated weights of 30 to punctuality, 60 to safety of cargo and 10 to convenience.

The manager then rated the performance of each form of transport on the different attributes. The values she assigned are shown below, together with the estimated annual cost of using each form of transport.

		Benefits		
Form of transport	Punctuality	Safety	Convenience	Costs ($)
Air	100	70	60	150 000
Sea	0	60	80	90 000
Road and Ferry	60	0	100	40 000
Rail and Ferry	70	100	0	70 000

(a) Determine the form of transport which has the highest valued overall benefits, assuming that mutual preference independence exists between the attributes.

(b) For each form of transport, plot the value of overall benefits against costs and hence identify the forms of transport which lie on the efficient frontier.

(c) If the manager would be prepared to pay $70 000 per year to move from the least safe to the most safe form of transport, determine which alternative she should select.

(5) A local authority has to decide on the location of a new waste-disposal facility and five sites are currently being considered; Inston Common, Jones Wood, Peterton, Red Beach and Treehome Valley. In order to help them to choose between the sites the managers involved in the decision arranged for a decision analyst to attend one of their meetings. He first got the managers to consider the factors which they thought were relevant to the decision and, after some debate, four factors were identified:

(i) The visual impact of the site on the local scenery (for example, a site at Treehome Valley would be visible from a nearby beauty spot);

(ii) The ease with which waste could be transported to the site (for example, Red Beach is only two miles from the main town in the area and is close to a main highway while Inston Common is in a remote spot and its use would lead to a major increase in the volume of transport using the minor roads in the area);

(iii) The risk that the use of the site would lead to contamination of the local environment (e.g. because of leakages of chemicals into watercourses);

(iv) The cost of developing the site.

The decision analyst then asked the managers to assign scores to the sites to show how well they performed on each of the first three attributes. The scores they eventually agreed are shown below, together with the estimated cost of developing each site. Note that 0 represents the worst and 100 the best score on an attribute. In the case of risk, therefore, a score of 100 means that a site is the least risky.

| Site | Benefits | | | Costs |
	Visual impact	Ease of transport	Risk	($ million)
Inston Common	100	0	60	35
Jones Wood	20	70	100	25
Peterton	80	40	0	17
Red Beach	20	100	30	12
Treehome Valley	0	70	60	20

The decision analyst then asked the managers to imagine a site which had the worst visual impact, the most difficult transport requirements and the highest level of risk. He then asked them, if they had a chance of switching from this site to one which had just one of the benefits at its best value, which would they choose? The managers agreed that they would move to a site offering the least risk of contamination. A move to a site with the best visual impact was considered to be 80% as preferable as this, while a move to one with the most convenient transport facilities was 70% as preferable.

(a) Can we conclude from the values which were assigned to the different sites for visual impact that, in terms of visual impact, the Inston Common site is five times preferable to Red Beach? If not, what can we infer from the figures?

(b) An alternative way of allocating weight to the three benefit attributes would have involved asking the managers to allocate a score reflecting the importance of each attribute. For example, they might have judged that risk was five times more important and visual impact three times more important than ease of transport, so that weights of 5, 3 and 1 would have been attached to the attributes. What are the dangers of this approach?

(c) Assuming that mutual preference independence exists between the attributes, determine the value of aggregate benefits for each site.

(d) Plot the aggregate benefits and costs of each site on a graph and hence identify the sites which lie on the efficient frontier.

(e) Although a weight of 80 was finally agreed for visual impact, this was only after much debate and some managers still felt that a weight of 65 should have been used while others thought that 95 would have been more appropriate. Perform a sensitivity analysis on the weight assigned to visual impact to examine its effect on the aggregate benefits of the sites and interpret your results.

REFERENCES

1. Wright, G. (1984) *Behavioral Decision Theory*, Sage, Beverly Hills, California.
2. Keeney, R. L. (1982) Decision Analysis: an Overview, *Operations Research*, **30**, No. 5, 803–837.
3. Keeney, R. L. and Raiffa, H. (1976) *Decisions With Multiple Objectives: Preferences and Value Tradeoffs*, Wiley, New York.
4. Edwards, W. (1971) Social Utilities, *Engineering Economist*, Summer Symposium Series, 6.
5. Watson, S. R. and Buede, D. M. (1987) *Decision Synthesis*, Cambridge University Press, Cambridge.
6. Wooler, S. and Barclay, S. (1988) Strategy for Reducing Dependence on a Strike-prone Production Facility, in P. Humphreys, A. Vari, J. Vecsenyi and O. Larichev (eds) *Strategic Decision Support Systems*, North-Holland, Amsterdam.

7. Buede, D. M. and Choisser, R. W. (1984) An Aid for Evaluators of System Design Alternatives, *Defense Management Journal*, 32–38.
8. Brownlow, S. A. and Watson, S. R. (1987) Structuring Multi-attribute Value Hierarchies, *Journal of the Operational Research Society*, **38**, 309–317.
9. Edwards, W. and Newman, J. R. (1986) Multiattribute Evaluation, in H. R. Arkes and K. R. Hammond (eds) *Judgment and Decision Making*, Cambridge University Press, Cambridge.
10. Von Winterfeldt, D. and Edwards, W. (1986) *Decision Analysis and Behavioral Research*, Cambridge University Press, Cambridge.
11. Bodily, S. E. (1985) *Modern Decision Making*, McGraw-Hill, New York.
12. Phillips, L. D. (1984) A Theory of Requisite Decision Models, *Acta Psychologica*, **56**, 29–48.
13. Barclay, S (1987) *A User's Manual to HIVIEW*, Decision Analysis Unit, London School of Economics.

Chapter 3

Introduction to Probability

INTRODUCTION

In the previous chapter we discussed the analysis of decisions where uncertainty was not considered to be a major factor. However, in many problems the decision maker is not sure what will happen if a particular course of action is chosen. A company that is considering the purchase of an expensive new machine will face uncertainty relating to factors such as the machine's reliability, life span and resale value. Similarly, an individual who has to decide whether or not to purchase household insurance will be uncertain as to whether his home will be burgled, flooded or damaged by fire. In the next chapter we will be looking at how to analyze decisions which involve uncertainty, but before this we need to consider how the concept of probability can be used to provide a measure of uncertainty.

There are a number of ways in which uncertainty can be measured and expressed. The simplest method involves the use of words such as 'unlikely', 'almost impossible', 'probable', 'doubtful' and 'expected'. Unfortunately, it has been found that different people attach very different meanings to these expressions, and even individuals are not consistent over time in their use of them. For example, Moore and Thomas[1] discuss an experiment at a business school where a large number of executives were asked to rank 10 words or phrases in decreasing order of uncertainty. The ranking of the word 'likely' ranged from second to seventh, while the ranking for 'unlikely' ranged from third to tenth.

Numbers offer a much more precise way of measuring uncertainty, and most people will be familiar with the use of odds. While many decision makers may be happy to use expressions such as '100 to 1 against' or 'evens', odds do have the disadvantage that they are awkward to handle arithmetically when, for example, we want to determine the chances that a number of different outcomes will occur.

Because of this, we will be using the concept of probability in our decision models. Probabilities are measured on a scale which runs from 0 to 1. If the probability of an outcome occurring is zero then this implies that the outcome is impossible. At the opposite extreme, if it is considered that an outcome is certain to occur then this will be represented by a probability of 1 and the greater the chances of the event occurring, the closer its probability will be to 1. It is worth pointing out that odds can be converted to probabilities. For example, odds of 50 to 1 against imply that there are 50 'chances' that the outcome will not occur and one chance that it will: a total of 51 'chances'. Thus the chances of the event occurring are, in probability terms, 1 in 51 (or 0.0196). 'Evens' is, of course, represented by the probability of 1/2.

In this chapter we will introduce the main ideas and rules which are used in probability calculations. The ways in which decision analysts elicit subjective probabilities from decision makers will be described and evaluated in Chapter 10. Throughout the book we will be using the notation $p(\ )$ to mean 'the probability of . . .'. For example, we will write 'the probability of rain' as $p(\text{rain})$.

OUTCOMES AND EVENTS

Before proceeding, we need to be clear in our definitions of outcomes and events. Suppose that a company is thinking of simultaneously launching two new products, A and B. The company's marketing manager decides to list all the possible things that can happen if the simultaneous launch goes ahead. His list is shown below:

Both products fail
Product A succeeds but B fails
Product A fails but B succeeds
Both products succeed.

Each of the four possible things that can happen is called an *outcome*. An *event* consists of one or more possible outcomes. For example, the event 'just one product succeeds' consists of the two outcomes: 'A succeeds but B fails' and 'A fails but B succeeds'. The event 'at least one product succeeds' consists of the last three outcomes in the list. However, the event 'both products fail' clearly consists of only one outcome.

APPROACHES TO PROBABILITY

There are three different approaches to deriving probabilities: the classical approach, the relative frequency approach and the subjective approach. The first two methods lead to what are often referred to as objective probabilities because, if they have access to the same information, different people using either of these approaches should arrive at exactly the same probabilities. In contrast, if the subjective approach is adopted it is likely that people will differ in the probabilities which they put forward.

The Classical Approach

Consider the following problem. You work for a company which is a rather dubious supplier of electronic components and you have just sent a batch of 200 components to a customer. You know that 80 of the components are damaged beyond repair, 30 are slightly damaged and the rest are in working order. Moreover, you know that before he signs the acceptance form the customer always picks out one component at random and tests it. What are the chances that the customer will select a component which is damaged beyond repair?

The classical approach to probability involves the application of the following formula:

The probability of an event occurring

$$= \frac{\text{Number of outcomes which represent the occurrence of the event}}{\text{Total number of possible outcomes}}$$

In our problem the customer could select any one of the 200 components, so there are 200 possible outcomes. In 80 of these outcomes a component is selected which is damaged beyond repair so:

p(selected component is damaged beyond repair)$=80/200=0.40$

In order to apply the classical approach to a problem we have to assume that each outcome is equally likely to occur, so in this case we would have to assume that the customer is equally likely to select each component. Of course, this would not be the case if you knew that the customer tended to select a component from the top of the box and you deliberately packed the defective components in the bottom. In most practical situations (e.g. the simultaneous product launch above) the outcomes will not be equally likely and therefore the usefulness of this approach is limited.

The Relative Frequency Approach

In the relative frequency approach the probability of an event occurring is regarded as the proportion of times that the event occurs in the long run if stable conditions apply. This probability can be estimated by repeating an experiment a large number of times or by gathering relevant data and determining the frequency with which the event of interest has occurred in the past. For example, a quality control inspector at a factory might test 250 light bulbs and find that only eight are defective. This would suggest that the probability of a bulb being defective is 8/250 (or 0.032). The reliability of the inspector's probability estimate would improve as he gathered more data: an estimate based on a sample of 10 bulbs would be less reliable than one based on the sample of 250. Of course, the estimate is only valid if manufacturing conditions remain unchanged. Similarly, if the publisher of a weekly magazine found that circulation had exceeded the break-even level in 35 out of the past 60 weeks then he might estimate that the probability of sales exceeding the break-even level next week is 35/60 (or 0.583). Clearly, for this probability estimate to be reliable the same market conditions would have to apply to every week under consideration; if there is a trend or seasonal pattern in sales it would not be reliable.

This raises the problem of specifying a suitable reference class. For example, suppose that we wish to determine the probability that Mary, a 40-year-old unemployed computer programmer, will find a job within the next 12 months. By looking at recent past records we might find that 30% of unemployed people found jobs within a year and hence estimate that the probability is 0.3. However, perhaps we should only look at those records relating to unemployed female computer programmers of Mary's age and living in Mary's region of the country, or perhaps we should go even further and only look at people with similar qualifications and take into account the fact that Mary has a record of ill health. Clearly, if the data we used were made too specific it is likely that we would find that the only relevant record we had related to Mary herself. It can be seen that there is a conflict between the desirability of having a large set of past data and the need to make sure that the data relate closely to the event under consideration. Judgment is therefore required to strike a balance between these two considerations.

The Subjective Approach

Most of the decision problems which we will consider in this book will require us to estimate the probability of unique events occurring (i.e. events which only occur once). For example, if a company needs to estimate the probability that a new product will be successful or that a new state-of-the-art machine will perform reliably, then, because of the uniqueness of the situation, the past data required by the relative frequency approach will not be available. The company may have access to data relating to the success or otherwise of earlier products or machines, but it unlikely that the conditions that applied in these past situations will be directly relevant to the current problem. In these circumstances the probability can be estimated by using the subjective approach. A subjective probability is an expression of an individual's degree of belief that a particular event will occur. Thus a sales manager may say: 'I estimate that there is a 0.75 probability that the sales of our new product will exceed $2 million next year.' Of course, such a statement may be influenced by past data or any other information which the manager considers to be relevant, but it is ultimately a personal judgment, and as such it is likely that individuals will differ in the estimates they put forward even if they have access to the same information.

Many people are sceptical about subjective probabilities and yet we make similar sorts of judgments all the time. If you decide to risk not insuring the contents of your house this year then you must have made some assessment of the chances of the contents remaining safe over the next 12 months. Similarly, if you decide to invest on the stock market, purchase a new car or move to a new house you must have spent some time weighing up the chances that things will go wrong or go well. In organizations, decisions relating to the appointment of new staff, launching an advertising campaign or changing to a new computer system will require some evaluation of the uncertainties involved. As we argued in the Introduction, by representing this judgment numerically rather than verbally a much less vague assessment is achieved. The resulting statement can be precisely communicated to others and, as we shall see in Chapter 11, it enables an individual's views to be challenged and explored.

Some people may be concerned that subjective probability estimates are likely to be of poor quality. Much research has been carried out by psychologists to find out how good people are at making these sorts of judgments. We will review this research in Chapters 8 and 9, while in Chapter 10 we will introduce a number of elicitation methods which are designed to help decision makers to make judgments about probabilities. At this stage, however, it is worth pointing out that such judgments rarely need to be exact. As we shall see, sensitivity analysis often reveals that quite major changes in the probabilities are required before it becomes apparent that the decision maker should switch from one course of action to another.

Having looked at the three approaches to probability, we now need to consider the concepts and rules which are used in probability calculations. These calculations apply equally well to classical, relative frequency or subjective probabilities.

MUTUALLY EXCLUSIVE AND EXHAUSTIVE EVENTS

Two events are mutually exclusive (or *disjoint*) if the occurrence of one of the events precludes the simultaneous occurrence of the other. For example, if the sales of a product in the USA next year exceed 10 000 units they cannot also be less than 10 000 units. Similarly, if a quality control inspection of a new TV set reveals that it is in perfect working order it cannot simultaneously be defective. However, the events of 'dollar rises against the yen tomorrow' and 'the Dow-Jones index falls tomorrow' are not mutually exclusive: there is clearly a possibility that both events can occur together. If you make a list of the events which can occur when you adopt a particular course of action then this list is said to be *exhaustive* if your list includes *every* possible event.

THE ADDITION RULE

In some problems we need to calculate the probability that *either* one event *or* another event will occur (if A and B are the two events, you may see 'A or B' referred to as the 'union' of A and B). For example, we may need to calculate the probability that a new product development will take either 3 or 4 years, or the probability that a construction project will be delayed by either bad weather or a strike. In these cases the addition rule can be used to calculate the required probability but, before applying the rule, it is essential to establish whether or not the two events are mutually exclusive.

If the events are mutually exclusive then the addition rule is:

$$p(A \text{ or } B) = p(A) + p(B) \qquad \text{(where A and B are the events)}$$

For example, suppose that a manager estimates the following probabilities for the time that a new product will take to launch:

Time to launch product	Probability
1 year	0.1
2 years	0.3
3 years	0.4
4 years	0.2

Suppose that we want to use this information to determine the probability that the launch will take either 1 or 2 years. Clearly, both events are mutually exclusive so:

$$p(\text{launch takes 1 or 2 years}) = p(\text{takes 1 year}) + p(\text{takes 2 years})$$
$$= 0.1 + 0.3 = 0.4$$

Similarly, if we want to determine the probability that the launch will take at least 2 years we have:

$$p(\text{launch takes 2 or 3 or 4 years}) = 0.3 + 0.4 + 0.2 = 0.9$$

Note that the complete set of probabilities given by the manager sum to 1, which implies that the list of possible launch times is exhaustive. In the manager's view it is certain that the launch will take 1, 2, 3 or 4 years.

Let us now see what happens if the addition rule for mutually exclusive events is wrongly applied. Consider Table 3.1. This relates to a tidal river which is liable to cause flooding during the month of April. The table gives details of rainfall during April for the past 20 years and also whether or not the river caused flooding. For example, there was light rainfall and yet the river flooded in four out of the last 20 Aprils.

Table 3.1 The frequency of flooding of a tidal river in April over the last 20 years

	Rainfall		(number of years)
	Light	Heavy	Total
River flooded	4	9	13
River did not flood	5	2	7
Total	9	11	20

Suppose that in order to make a particular decision we need to calculate the probability that next year there will either be heavy rain or the river will flood. We decide to use the relative frequency approach based on the records for the past 20 years and we then proceed as follows:

$$p(\text{heavy rain or flood}) = p(\text{heavy rain}) + p(\text{flood})$$
$$= 11/20 + 13/20 = 24/20 \text{ which exceeds one!}$$

The mistake we have made is to ignore the fact that heavy rain and flooding are *not* mutually exclusive: they can and have occurred together. This has meant that we have double-counted the nine years when both events did occur, counting them both as heavy rain years and as flood years.

If the events are not mutually exclusive we should apply the addition rule as follows:

$$p(A \text{ or } B) = p(A) + p(B) - p(A \text{ and } B)$$

The last term has the effect of negating the double-counting. Thus the correct answer to our problem is:

$$p(\text{heavy rain or flood}) = p(\text{heavy rain}) + p(\text{flood}) - p(\text{heavy rain and flood})$$
$$= 11/20 + 13/20 - 9/20 = 15/20 \text{ (or 0.75)}$$

COMPLEMENTARY EVENTS

If A is an event then the event 'A does not occur' is said to be the complement of A. For example, the complement of the event 'project completed on time' is the event 'project not completed on time', while the complement of the event 'inflation exceeds 5% next year' is the event 'inflation is less than or equal to 5% next year'. The complement of event A can be written as $\bar{A}$ (pronounced 'A bar').

Since it is certain that either the event or its complement must occur their probabilities always sum to one. This leads to the useful expression:

$$p(\bar{A}) = 1 - p(A)$$

For example, if the probability of a project being completed on time is 0.6, what is the probability that it will not be completed on time? The answer is easily found:

$$p(\text{not completed on time}) = 1 - p(\text{completed on time})$$
$$= 1 - 0.6 = 0.4$$

MARGINAL AND CONDITIONAL PROBABILITIES

Consider Table 3.2, which shows the results of a survey of 1000 workers who were employed in a branch of the chemicals industry. The workers have been classified on the basis of whether or not they were exposed in the past to a hazardous chemical and whether or not they have subsequently contracted cancer.

Table 3.2 Results of a survey of workers in a branch of the chemicals industry

| | Number of workers | | |
	Contracted cancer	Have not contracted cancer	Total
Exposed to chemical	220	135	355
Not exposed to chemical	48	597	645
Total	268	732	1000

Suppose that we want to determine the probability that a worker in this industry will contract cancer *irrespective* of whether or not he or she was exposed to the chemical. Assuming that the survey is representative and using the relative frequency approach, we have:

$$p(\text{worker contracts cancer}) = 268/1000 = 0.268$$

This probability is called an unconditional or *marginal* probability because it is not conditional on whether or not the worker was exposed to the chemical (note that it is calculated by taking the number of workers in the margin of the table).

Suppose that now we wish to calculate the probability of a worker suffering from cancer *given that* he or she was exposed to the chemical. The required probability is known as a *conditional probability* because the probability we are calculating is conditional on the fact that the worker has been exposed to the chemical. The probability of event A occurring given that event B has occurred is normally written as $p(A|B)$, so in our case we wish to find: $p(\text{worker contracts cancer}|\text{exposed to chemical})$. We only have 355 records of workers who were exposed to the chemical and *of these* 220 have contracted cancer, so:

$$p(\text{worker contracts cancer}|\text{exposed to chemical}) = 220/355 = 0.620$$

Note that this conditional probability is greater than the marginal probability of a worker contracting cancer (0.268), which implies that exposure to the chemical increases a worker's chances of developing cancer. We will consider this sort of relationship between events next.

INDEPENDENT AND DEPENDENT EVENTS

Two events, A and B, are said to be *independent* if the probability of event A occurring is unaffected by the occurrence or non-occurrence of event B. For example, the probability of a randomly selected husband belonging to blood group O will presumably be unaffected by the fact that his wife is blood group O (unless like blood groups attract or repel!). Similarly, the probability of very high temperatures occurring in England next August will not be affected by whether or not planning permission is granted next week for the construction of a new swimming pool at a seaside resort. If two events, A and B, are independent then clearly:

$$p(A|B) = p(A)$$

because the fact that B has occurred does not change the probability of A occurring. In other words, the conditional probability is the same as the marginal probability.

In the previous section we saw that the probability of a worker contracting cancer *was* affected by whether or not he or she has been exposed to a chemical. These two events are therefore said to be *dependent*.

THE MULTIPLICATION RULE

We saw earlier that the probability of either event A or B occurring can be calculated by using the addition rule. In many circumstances, however, we need to calculate the probability that *both* A *and* B will occur. For example, what is the probability that both the New York and the London Stock Market indices will fall today, or what is the probability that we will suffer strikes this month at both of our two production plants? The probability of A and B occurring is known as a *joint probability*, and joint probabilities can be calculated by using the multiplication rule.

Before applying this rule we need to establish whether or not the two events are independent. If they are, then the multiplication rule is:

$$p(A \text{ and } B) = p(A) \times p(B)$$

For example, suppose that a large civil engineering company is involved in two major projects: the construction of a bridge in South America and of a dam in Europe. It is estimated that the probability that the bridge construction will be completed on time is 0.8, while the probability that the dam will be completed on time is 0.6. The teams involved with the two projects operate totally indepedently, and the company wants to determine the probability that both projects will be completed on time.

Since it seems reasonable to assume that the two completion times are independent, we have:

$$p(\text{bridge and dam completed on time}) = p(\text{bridge completed on time})$$
$$\times p(\text{dam completed on time})$$
$$= 0.8 \times 0.6 = 0.48$$

The use of the above multiplication rule is not limited to two independent events. For example, if we have four independent events, A, B, C and D then:

$$p(A \text{ and } B \text{ and } C \text{ and } D) = p(A) \times p(B) \times p(C) \times p(D)$$

If the events are not independent the multiplication rule is:

$$p(A \text{ and } B) = p(A) \times p(B|A)$$

because A's occurrence would affect B's probability of occurrence. Thus we have the probability of A occurring multiplied by the probability of B occurring, given that A has occurred.

To see how the rule can be applied, consider the following problem. A new product is to be test marketed in Florida and it is estimated that there is a probability of 0.7 that the test marketing will be a success. If the test marketing is successful, it is estimated that there is a 0.85 probability that the product will be a success nationally. What is the probability that the product will be both a success in the test marketing and a success nationally?

Clearly, it is to be expected the probability of the product being a success nationally will depend upon whether it is successful in Florida. Applying the multiplication rule we have:

p(success in Florida and success nationally)
$$= p(\text{success in Florida}) \times p(\text{success nationally}|\text{success in Florida})$$
$$= 0.7 \times 0.85 = 0.595$$

PROBABILITY TREES

As you have probably gathered by now, probability calculations require clear thinking. One device which can prove to be particularly useful when awkward problems need to be solved is the probability tree, and the following problem is designed to illustrate its use.

A large multinational company is concerned that some of its assets in an Asian country may be nationalized after that country's next election. It is estimated that there is a 0.6 probability that the Socialist Party will win the next election and a 0.4 probability that the Conservative Party will win. If the Socialist Party wins then it is estimated that there is a 0.8 probability that the assets will be nationalized, while the probability of the Conservatives nationalizing the assets is thought to be only 0.3. The company wants to estimate the probability that their assets will be nationalized after the election.

The probability tree for this problem is shown in Figure 3.1. Note that the tree shows the possible events in chronological order from left to right; we consider first which party will win the election and then whether each party will or will not nationalize the assets. The four routes through the tree represent the four joint events which can occur (e.g. Socialists win and assets are not nationalized). The calculations shown on the tree are explained below.

We first determine the probability that the Socialists will win and the assets will be nationalized using the multiplication rule for dependent events:

p(Socialists win and assets nationalized)
$$= p(\text{Socialists win}) \times p(\text{assets nationalized}|\text{Socialists win})$$
$$= 0.6 \times 0.8 = 0.48$$

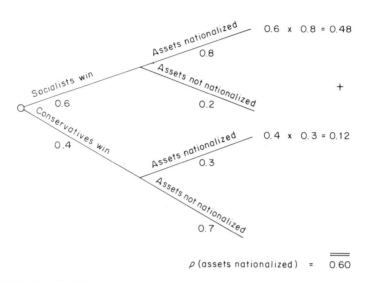

Figure 3.1 A probability tree

We then determine the probability that the Conservatives will win and that the assets will be nationalized:

p(Conservatives win and assets nationalized)
$\quad = p$(Conservatives win)$\times p$(assets nationalized|Conservatives win)
$\quad = 0.4 \times 0.3 = 0.12$

Now we can obtain the overall probability of the assets being nationalized as follows:

p(assets nationalized)
$= p$(*either* Socialists win and nationalize *or* Conservatives win and nationalize)

These two events are mutually exclusive, since we assume that the election of one party precludes the election of the other, so we can simply add the two probabilities we have calculated to get:

$$p\text{(assets nationalized)} = 0.48 + 0.12 = 0.60$$

PROBABILITY DISTRIBUTIONS

So far in this chapter we have looked at how to calculate the probability that a *particular* event will occur. However, when we are faced with a decision

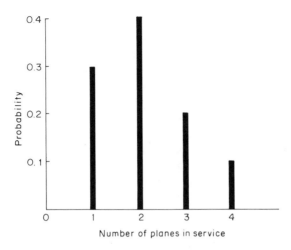

Figure 3.2 Probability distribution for number of planes in service by end of year

it is more likely that we will be concerned to identify all the possible events which could occur, if a particular course of action was chosen, together with their probabilities of occurrence. This complete statement of all the possible events and their probabilities is known as a probability distribution. For example, a consortium of business people who are considering setting up a new airline might estimate the following probability distribution for the number of planes they will be able to have in service by the end of the year (this distribution is illustrated in Figure 3.2):

No. of planes in service	Probability
1	0.3
2	0.4
3	0.2
4	0.1
	1.0

Note that the probabilities sum to one, since all the possible events have been listed. The 'number of planes in service' is known as an uncertain quantity. If we plotted, on a continuous scale, the values which this quantity could assume then there would be gaps between the points: it would clearly be impossible for the airline to have 2.32 or 3.2451 planes in service since the number of planes must be a whole number. This is therefore an example of what is known as a *discrete* probability distribution.

  In contrast, in a *continuous* probability distribution the uncertain quantity can take on any value within a specified interval. For example, the time taken to assemble a component on a production line could take on any value between, say, 0 and 30 minutes. There is no reason why the time should be restricted to a whole number of minutes. Indeed, we might wish to express it in thousandths or even millionths of a minute; the only limitation would be the precision of our measuring instruments.

Because continuous uncertain quantities can, in theory, assume an infinite number of values, we do not think in terms of the probability of a particular value occurring. Instead, the probability that the variable will take on a value within a given range is determined (e.g. what is the probability that our market share in a year's time will be between 5% and 10%?). Figure 3.3 shows a probability distribution for the time to complete a construction project. Note that the vertical axis of the graph has been labelled *probability density* rather than probability because we are not using the graph to display the probability that exact values will occur. The curve shown is known as a *probability density function* (pdf). The probability that the completion time will be between two values is found by considering the *area* under the pdf between these two points. Since the company is certain that the completion time will be between 10 and 22 weeks, the whole area under the curve is equal to 1. Because half of the area under the curve falls between times of 14 and 18 weeks this implies that there is a 0.5 probability that the completion time will be between these two values. Similarly, 0.2 (or 20%) of the total area under the curve falls between 10 and 14 weeks, implying that the probability that the completion time will fall within this interval is 0.2. A summary of the probability distribution is shown below.

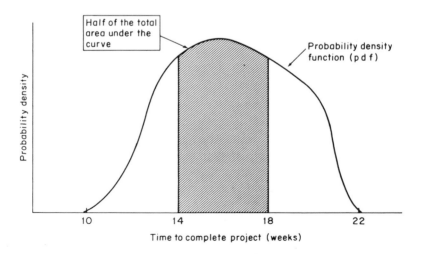

Figure 3.3 Probability distribution for project-completion time

Project completion time	Probability
10 to under 14 weeks	0.2
14 to under 18 weeks	0.5
18 to under 22 weeks	0.3
	1.0

When eliciting a probability distribution it is sometimes easier to think in terms of the probability of a variable having a value less than a particular figure. For example, 'what is the probability that our market share in a year's time will be less than 10%?' This can be facilitated by deriving the *cumulative distribution function* (cdf), which gives the probability that a variable will have a value less than a particular value. The cdf for the above project is shown in Figure 3.4. It can be seen that there is a 0.2 probability that the completion time will be less than 14 weeks, a 0.7 probability that it will be less than 18 weeks and it is certain that the time will be less than 22 weeks.

Sometimes it is useful to use continuous distributions as approximations for discrete distributions and vice versa. For example, when a discrete variable can assume a large number of possible values it may be easier to treat it as a continuous variable. In practice, monetary values can usually be regarded as continuous because of the very large number of values which can be assumed within a specified range (consider, for example, the possible

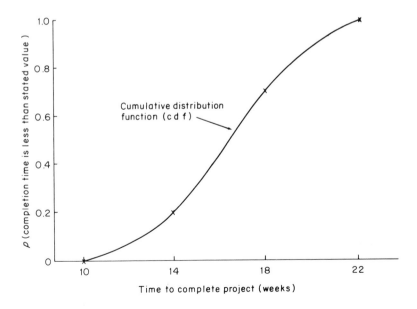

Figure 3.4 Cumulative distribution function for project-completion time

revenues in the range $0 to $1 million which could be earned by a company). This might also apply to the sales of a product. For example, the number of tins of baked beans sold must be an integer, but if sales can range from 0 to 5 million tins then, again, the uncertain quantity can take on a very large number of possible values. Similarly, it is often convenient to use discrete distributions as approximations to continuous distributions, particularly when constructing decision tree models (see Chapter 5). For example, we might approximate the continuous distribution of project completion times above by using the midpoints of the three intervals to obtain the following discrete distribution:

Project completion time	Probability
12 weeks	0.2
16 weeks	0.5
20 weeks	0.3
	___
	1.0

EXPECTED VALUES

Suppose that a retailer runs a small shop selling television sets. The number of color sets she sells per week follows the probability distribution shown below:

No. of sets sold	Probability
0	0.01
1	0.10
2	0.40
3	0.30
4	0.10
5	0.09
	____
	1.00

If this probability distribution applies to all weeks (e.g. we will assume there is no trend or seasonal pattern in her sales) then we might be interested in calculating her mean (or average) weekly sales. This is easily done by multiplying each sales level by its probability of occurrence and summing the resulting products as shown below. The result is known as an expected value.

No. of sets sold	Probability	No. of sets × probability
0	0.01	0
1	0.10	0.10
2	0.40	0.80
3	0.30	0.90
4	0.10	0.40
5	0.09	0.45
	1.00	Expected sales = 2.65

It can be seen that an expected value is a weighted average with each possible value of the uncertain quantity being weighted by its probability of occurrence. The resulting figure represents the mean level of sales which would be expected if we looked at the sales records over a large number of weeks. Note that an expected value does not have to coincide with an actual value in the distribution; it is obviously not possible to sell 2.65 sets in a given week.

Although an expected value is most easily interpreted as 'an average value which will result if a process is repeated a large number of times', as we will see in the next chapter, we may wish to use expected values even in unique situations. For example, suppose that a company purchases its main raw material from a country which has just experienced a military coup. As a result of the coup, it is thought that there is some possibility that the price of the material will increase in the very near future, and the company is therefore thinking of purchasing a large supply of the material now. It estimates that there is a 0.7 probability that the price will increase, in which case a saving of $350 000 will be made, and a 0.3 probability that the price will fall because of other world market conditions. In this case, purchasing early will have cost $200 000. What is the expected saving of purchasing early? The calculations are shown below:

$$\text{Expected savings} = (0.7 \times \$350\,000) + (0.3 \times -\$200\,000)$$
$$= \$185\,000$$

Note that savings of $185 000 will not be achieved; the company will either save $350 000 or lose $200 000. The figure is simply an average of the two monetary values taking into account their probabilities of occurrence. The practical use of such a figure is that it allows a decision maker to evaluate the attractiveness of different options in decision problems which involve uncertainty, the topic of the next chapter.

THE AXIOMS OF PROBABILITY THEORY

If you use subjective probabilities to express your degree of belief that events will occur then your thinking must conform to the axioms of probability theory. These axioms have been implied by the preceding discussion, but we will formally state them below.

Axiom 1: Positiveness
The probability of an event occurring must be non-negative.
Axiom 2: Certainty
The probability of an event which is certain to occur is 1. Thus axioms 1 and 2 imply that the probability of an event occurring must be at least zero and no greater than 1.
Axiom 3: Unions
If events A and B are mutually exclusive then:

$$p(A \text{ or } B) = p(A) + p(B)$$

It can be shown that all the laws of probability that we have considered in this chapter can be derived from these three axioms. Note that they are generally referred to as Kolmogoroff's axioms and, as stated above, they relate to situations where the number of possible outcomes is finite.

In the next few chapters we will use subjective probability assessments in our calculations without attempting to evaluate the quality of these judgmental inputs to our analyses. In Chapters 8–10 we will consider the degree to which probability judgments comply with the axioms and have validity as predictions of future events.

SUMMARY

As we shall see in the next chapter, probability assessments are a key element of decision models when a decision maker faces risk and uncertainty. In most practical problems the probabilities used will be subjective, but they must still conform to the underlying axioms of probability theory. Again, our approach has been normative; probability calculus is designed to show you what your judgments should look like if you accept its axioms and think rationally.

In later chapters we will look at methods which are designed to help the decision maker to generate coherent assessments, and we will examine in detail how good individuals are at making judgments about probabilities. Often the receipt of new information, such as market research results or provisional sales figures, can be used to modify initial probability assessments, and in Chapter 7 we will show how this revision of opinion should be carried out.

Above all, the correct application of the rules and concepts which we have introduced in this chapter requires both practice and clarity of thought. You are therefore urged to attempt the following exercises before reading further.

EXERCISES

(1) Determine the probability of each of the following events occurring. State the approach to probability which you used and any assumptions which you needed to make.
   (a) A person selected from the payroll of a company is a clerical worker given that there are 350 people on the payroll of whom 120 are clerical workers.
   (b) A light bulb selected from a production line is defective if, out of 400 bulbs already tested, eight were defective.
   (c) A new-born baby is male.
   (d) This month's sales forecast for a product has an error of more than 10% if the forecast had an error of over 10% in 21 out of the last 60 months.
   (e) A permanently manned space station is established on Mars by the year 2050.
(2) The following table shows an estimated probability distribution for the sales of a new product in its first week:

Number of units sold	0	1	2	3	4	5
Probability	0.05	0.15	0.20	0.35	0.15	0.10

What is the probability that in the first week:
   (a) Four or five units will be sold;
   (b) At least 3 units will be sold;
   (c) At least 1 unit will be sold?
(3) The managers of a food company are interested in determining the effect on their sales of a competitor's television advertisements. An analysis of sales records for the last 120 weeks gives the following results:

| | Level of sales | | | (no. of weeks) |
	Low	Medium	High	Total
Competitor advertised	32	14	18	64
Competitor did not advertise	21	12	23	56
Total	53	26	41	120

Assuming that these past data are a reliable guide to the future, determine the probability that next week:

(a) The competitor will advertise;

(b) Sales will not be high;

(c) Medium or high sales will be achieved;

(d) Either the competitor will advertise or only low sales will be achieved;

(e) Either the competitor will not advertise or high sales will be achieved.

(4) (a) With reference to the table in question 3, determine the following probabilities:

    (i) p(next week's sales will be high)

    (ii) p(next week's sales will be high|the competitor advertises)

    (iii) p(next week's sales will be high|the competitor does not advertise)

    (iv) p(next week's sales will be low)

    (v) p(next week's sales will be low|the competitor advertises)

(b) Do the events 'competitor advertises' and 'high sales' appear to be independent?

(5) Given below are the results of a survey of 100 cars:

| | Condition of brakes | | |
Condition of tyres	Faulty	Not faulty	Total
Faulty	25	5	30
Not faulty	15	55	70
Total	40	60	100

(a) Assuming that the survey is representative of all the cars on the road, what is the probability that a car selected at random will have:

    (i) Faulty brakes;

    (ii) Faulty tyres;

    (iii) Either faulty brakes or faulty tyres;

    (iv) Faulty brakes given that it has faulty tyres;

    (v) Faulty tyres given that it has faulty brakes?

(b) What conclusion would you draw about the relationship between the events 'faulty tyres' and 'faulty brakes'?

(6) Three machines, A, B and C, operate independently in a factory. Machine A is out of action for 10% of the time, while B is out of action for 5% of the time and C for 20% of the time. A rush order has to be commenced at midday tomorrow. What is the probability that at this time:

  (a) All three machines will be out of action?

  (b) None of the machines will be out of action?

(7) A speculator purchases three stocks on the London Stock Exchange. He estimates that the probabilities that each of these stocks will have risen in value by the end of the week are, respectively, 0.6, 0.8 and 0.4.

  (a) Assuming that the price changes in the three stocks are independent, determine the probability that all three stocks will have risen in value by the end of the week.

  (b) Do you think that it is reasonable to assume that movements in the prices of individual stocks are independent?

(8) The managers of a company are considering the launch of a new product and they are currently awaiting the results of a market research study. It is thought that there is a 0.6 probability that the market research will indicate that sales of the product in its first three years will be high. If this indication is received then the probability that sales will be high is thought to be 0.8. What is the probability that the market research will indicate high sales and sales will turn out to be high?

(9) An engineer at a chemical plant wishes to assess the probability of a major catastrophe occurring at the plant during the overhaul of a processor as a result of a malfunction in the equipment being used in the overhaul. He estimates that the probability of a malfunction occurring is 0.1. If this happens there is only a 0.001 probability that a safety device will fail to switch off the equipment. If the safety device fails the probability of a major catastrophe is estimated to be 0.8. What is the probability that a major catastrophe will occur at the plant during the overhaul as a result of the equipment malfunctioning?

(10) The probability of the Dow-Jones index rising on the first day of trading next week is thought to be 0.6. If it does rise then the probability that the value of shares in a publishing company will rise is 0.8. If the index does not rise then the publishing company's shares will only have a 0.3 probability of rising. What is the probability that the publishing company's shares will rise in value on the first day of trading next week?

(11) A company has two warehouses to sell, one in the town of Kingstones and another in the nearby suburb of Eadleton. Because there have been some indications that property prices are likely to rise over the next six months some of the managers of the company are urging that the sale should be delayed. It is thought that the probability of the Kingstones warehouse rising in value by at least 10% is 0.6. If it does rise in value by this amount then there is a 0.9 probability that the Eadleton property will also rise in value by at least 10%. However, if the Kingstones

property rises in value by less than 10% there is only a 0.3 probability that the Eadleton warehouse will increase its value by at least 10%. What is the probability that (a) both warehouses; (b) only one warehouse will increase in value by at least 10% over the 6-month period?

(12) A car owners' club which offers a rescue service for stranded motorists has to make a decision on the number of breakdown patrols to deploy between midnight and 8 a.m. on weekdays during the summer. The number of requests for assistance received by a local office during these hours follows the probability distribution shown below.

No of requests received	0	1	2	3	4	5
Probability	0.01	0.12	0.25	0.42	0.12	0.08

(a) Calculate the expected number of requests received and interpret your result.

(b) Is this a discrete or a continuous probability distribution?

(13) You are thinking of selling your house and you reckon that there is a 0.1 probability that you will sell it for $120 000, a 0.5 probability that you will receive $100 000 for it and a 0.4 probability that you will only receive $80 000. What is the expected selling price of the property? Interpret your result.

(14) A toll bridge over the River Jay is operated by a private company who are thinking of installing automatic machines to collect the tolls. These machines, however, are not perfectly reliable and it is thought that the number of breakdowns occurring per day would follow the probability distribution shown below:

Number of breakdowns per day	0	1
Probability	0.4	0.6

When a breakdown occurred, revenue from tolls would be lost until the equipment was repaired. Given below are approximate probability distributions for the equipment repair time and the average revenue lost per hour.

Equipment repair time	Probability	Average revenue lost per hour	Probability
1 hour	0.7	$40	0.6
2 hours	0.3	$50	0.3
		$60	0.1

(a) Determine the probability distribution of revenue which would be lost per day as a result of machine breakdowns (it can be assumed that the above probability distributions are independent).
(b) Calculate the expected loss of revenue per day and interpret your result.

REFERENCE

1. Moore, P. G. and Thomas, H. (1988) *The Anatomy of Decisions* (2nd edn), Penguin, Harmondsworth.

Chapter 4

Decision Making under Uncertainty

INTRODUCTION

In many decisions the consequences of the alternative courses of action cannot be predicted with certainty. A company which is considering the launch of a new product will be uncertain about how successful the product will be, while an investor in the stock market will generally be unsure about the returns which will be generated if a particular investment is chosen. In this chapter we will show how the ideas about probability, which we introduced in Chapter 3, can be applied to problems where a decision has to be made under conditions of uncertainty.

We will first consider a method which is based on the expected value concept that we met in Chapter 3. Because an expected value can be regarded as an average outcome if a process is repeated a large number of times, this approach is arguably most relevant to situations where a decision is made repeatedly over a long period. A daily decision by a retailer on how many items to have available for sale might be an example of this sort of decision problem. In many situations, however, the decision is not made repeatedly, and the decision maker may only have one opportunity to choose the best course of action. If things go wrong then there will be no chance of recovering losses in future repetitions of the decision. In these circumstances some people might prefer the least risky course of action, and we will discuss how a decision maker's attitude to risk can be assessed and incorporated into a decision model.

Finally, we will broaden the discussion to consider problems which involve both uncertainty and more than one objective. As we saw in Chapter 2, problems involving multiple objectives are often too large for a decision maker to comprehend in their entirety. We will therefore look at a method which is designed to allow the problem to be broken down into smaller parts so that the judgmental task of the decision maker is made more tractable.

THE EXPECTED MONETARY VALUE (EMV) CRITERION

Consider the following problem. Each morning a food manufacturer has to make a decision on the number of batches of a perishable product which should be produced. Each batch produced costs $800 and each batch sold earns revenue of $1000. However, the daily demand for the product varies and follows the probability distribution shown below. Any batch which is unsold at the end of the day is worthless.

Demand per day (no. of batches)	1	2
Probability	0.3	0.7

The manufacturer would like to determine the optimum number of batches which he should produce each morning.

Clearly, the manufacturer has a dilemma. If he produces too many batches he will have wasted money in producing food which has to be destroyed at the end of the day. If he produces too few, he will be forgoing potential profits. We can represent his problem in the form of a *decision table* (Table 4.1). The rows of this table represent the alternative courses of action which are open to the decision maker (i.e. produce one or two batches), while the columns represent the possible levels of demand which are, of course, outside the control of the decision maker. The monetary values in the table show the profits which would be earned per day for the different levels of output and demand. For example, if one batch is produced and one batch demanded, a profit of $1000 – $800 (i.e. $200) will be made. This profit would also apply if two batches were demanded, since a profit can only be made on the batch produced.

Table 4.1 A decision table for the food manufacturer

(Daily profits)		Demand (no. of batches)	
		1	2
	Probability	0.3	0.7
Course of action			
Produce 1 batch		$200	$200
Produce 2 batches		– $600	$400

Given these potential profits and losses, how should the manufacturer make his decision? We will assume that he has only one objective, namely maximizing monetary gain so that other possible objectives such as maintaining customer goodwill or maximizing market share are of no concern to him. Moreover, this is clearly a decision which he faces every day, so the decision will be repeated a large number of times. In these circumstances it may be appropriate for him to choose the alternative which will lead to the highest *expected* daily profit. If he makes his decision on this basis then he is said to be using the *expected monetary value* or EMV criterion. You will recall from Chapter 3 that an expected value can be regarded as an average result which is obtained if a process is repeated a large number of times. It is calculated by multiplying each outcome by its probability of occurrence and then summing the resulting products. Thus the expected daily profits for the two production levels are:

Produce one batch:
expected daily profit $= (0.3 \times \$200) \quad + (0.7 \times \$200) = \$200$

Produce two batches:
expected daily profit $= (0.3 \times -\$600) + (0.7 \times \$400) = \$100$

These expected profits show that, in the long run, the highest average daily profit will be achieved by producing just one batch per day and, if the EMV criterion is acceptable to the food manufacturer, then this is what he should do.

Of course, the probabilities and profits used in this problem may only be rough estimates or, if they are based on reliable past data, they may be subject to change. We should therefore carry out sensitivity analysis to determine how large a change there would need to be in these values before the alternative course of action would be preferred. To illustrate the process, Figure 4.1 shows the results of a sensitivity analysis on the probability that just one batch will be demanded. Producing one batch will always yield an expected profit of $200, whatever this probability is. However, if the probability of just one batch being demanded is zero, then the expected profit of producing two batches will be $400. At the other extreme, if the probability of just one batch being demanded is 1.0 then producing two batches will yield an expected profit of $-$600. The line joining these points shows the expected profits for all the intermediate probabilities. It can be seen that producing one batch will continue to yield the highest expected profit as long as the probability of just one batch being demanded is greater than 0.2. Since currently this probability is estimated to be 0.3, it would take only a small change in the estimate for the alternative course of action to be preferred. Therefore in this case the probability needs to be estimated with care.

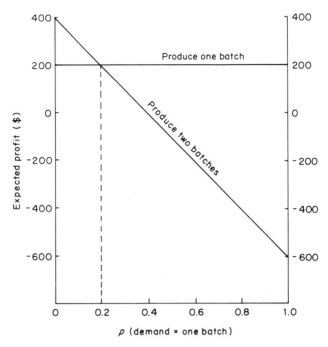

Figure 4.1 A sensitivity analysis for the food manufacturer's problem

LIMITATIONS OF THE EMV CRITERION

The EMV criterion may have been appropriate for the food manufacturer because he was only concerned with monetary rewards, and his decision was repeated a large number of times so that a long-run average result would have been of relevance to him. Let us now consider a different decision problem.

Imagine that you own a high-technology company which has been given the task of developing a new component for a large engineering corporation. Two alternative, but untried, designs are being considered (for simplicity, we will refer to these as designs 1 and 2), and because of time and resource constraints only one design can be developed. Table 4.2 shows the estimated net returns which will accrue to your company if each design is developed. Note that these returns depend on how successful the design is. The estimated probabilities of failure, partial success and total success for each design are also shown in the table.

The expected returns for design 1 are:

$$0.1 \times (-\$1m) + 0.1 \times \$0 + 0.8 \times (\$3m) = \$2.3m$$

Table 4.2 Returns and probabilities for the new component problem

| Course of action | Outcome | | | | | |
| | Total failure | | Partial success | | Total success | |
	Returns ($m)	Probability	Returns ($m)	Probability	Returns ($m)	Probability
Choose design 1	−1	0.1	0	0.1	3	0.8
Choose design 2	−6	0.3	1	0.1	10	0.6

while for design 2 the expected returns are:

$$0.3 \times (-\$6m) + 0.1 \times (\$1m) + 0.6 \times (\$10m) = \$4.3m$$

Thus according to the EMV criterion you should develop design 2, but would this really be your preferred course of action? There is a 30% chance that design 2 will fail and lead to a loss of $6 million. If your company is a small one or facing financial problems then these sort of losses might put you out of business. Design 1 has a smaller chance of failure, and if failure does occur then the losses are also smaller. Remember that this is a one-off decision, and there is therefore no chance of recouping losses on subsequent repetitions of the decision. Clearly, the risks of design 2 would deter many people. The EMV criterion therefore fails to take into account the attitude to risk of the decision maker.

This can also be seen in the famous St Petersburg paradox described by Bernoulli. Imagine that you are offered the following gamble. A fair coin is to be tossed until a head appears for the first time. If the head appears on the first throw you will be paid $2, if it appears on the second throw, $4, if it appears on the third throw $8 and so on. How much would you be prepared to pay to have the chance of engaging in this gamble? The expected returns on the gamble are:

$$\$2 \times (0.5) + \$4 \times (0.25) + \$8 \times (0.125) + \ldots \text{ etc.}$$
$$\text{which equals } 1 + 1 + 1 + \ldots \text{ to infinity}$$

so your expected returns will be infinitely large. On this basis, according to the EMV criterion, you should be prepared to pay a limitless sum of money to take part in the gamble. Given that there is a 50% chance that your return will be only $2 (and an 87.5% chance that it will be $8 or less), it is unlikely that many people would be prepared to pay anywhere near the amount prescribed by the EMV criterion!

It should also be noted that the EMV criterion assumes that the decision maker has a linear value function for money. An increase in returns from $0 to $1 million may be regarded by the decision maker as much more preferable than an increase from $9 million to $10 million, yet the EMV criterion assumes that both increases are equally desirable.

A further limitation of the EMV criterion is that it focuses on only one attribute: money. In choosing the design in the problem we considered above we may also wish to consider attributes such as the effect on company image of successfully developing a sophisticated new design, the spin-offs of enhanced skills and knowledge resulting from the development and the time it would take to develop the designs. All these attributes, like the monetary returns, would probably have some risk associated with them.

In the rest of this chapter we will address these limitations of the EMV criterion. First, we will look at how the concept of single-attribute utility can be used to take into account the decision maker's attitude to risk (or risk preference) in problems where there is just one attribute. The approach which we will adopt is based on the theory of utility which was developed by von Neumann and Morgenstern.[1] Then we will consider multi-attribute utility which can be applied to decision problems which involve both uncertainty and more than one attribute.

However, before we leave this section we should point out that the EMV criterion is very widely used in practice. Many people would argue that it is even appropriate to apply it to one-off decisions since, although an individual decision may be unique, a decision maker may, over time, make a large number of decisions that involve similar monetary sums so that returns will still be maximized by the consistent application of this criterion. Moreover, large organizations may be able to sustain losses on projects that represent only a small part of their operations. In these circumstances it may be reasonable to assume that risk neutrality applies, in which case the EMV criterion will be appropriate.

SINGLE-ATTRIBUTE UTILITY

The attitude to risk of a decision maker can be assessed by eliciting a *utility function*. This is to be distinguished from the value functions we met in Chapter 2. Value functions are used in decisions where uncertainty is not a major concern, and therefore they do not involve any consideration of risk attitudes. (We will, however, have more to say about the distinction between utility and value from a practical point of view in a later section of this chapter.)

To illustrate how a utility function can be derived, consider the following problem. A business woman who is organizing a business equipment exhibition in a provincial town has to choose between two venues: the Luxuria Hotel and the Maxima Center. To simplify her problem, she decides to estimate her potential profit at these locations on the basis of two scenarios:

high attendance and low attendance at the exhibition. If she chooses the Luxuria Hotel, she reckons that she has a 60% chance of achieving a high attendance and hence a profit of $30 000 (after taking into account the costs of advertising, hiring the venue, etc.). There is, however, a 40% chance that attendance will be low, in which case her profit will be just $11 000. If she chooses the Maxima Center, she reckons she has a 50% chance of high attendance, leading to a profit of $60 000, and a 50% chance of low attendance leading to a loss of $10 000.

We can represent the business woman's problem in the form of a diagram known as a decision tree (Figure 4.2). In this diagram a square represents a decision point; immediately beyond this, the decision maker can choose which route to follow. A circle represents a chance node. Immediately beyond this, chance determines, with the indicated probabilities, which route will be followed, so the choice of route is beyond the control of the decision maker. (We will consider decision trees in much more detail in Chapter 5.) The monetary values show the profits earned by the business woman if a given course of action is chosen and a given outcome occurs.

Now, if we apply the EMV criterion to the decision we find that the business woman's expected profit is $22 400 (i.e. 0.6×$30 000+0.4×$11 000) if she chooses the Luxuria Hotel and $25 000 if she chooses the Maxima Center. This suggests that she should choose the Maxima Center, but this

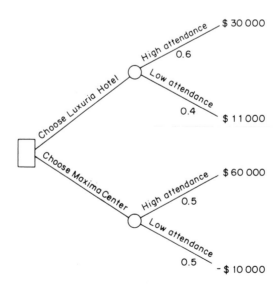

Figure 4.2 A decision tree for the conference organizer's problem

is the riskiest option, offering high rewards if things go well but losses if things go badly.

Let us now try to derive a utility function to represent the business woman's attitude to risk. We will use the notation $u(\ )$ to represent the utility of the sum of money which appears in the parentheses. First, we rank all the monetary returns which appear on the tree from best to worst and assign a utility of 1.0 to the best sum of money and 0 to the worst sum. As was the case with value functions in Chapter 2, any two numbers could have been assigned here as long as the best outcome is assigned the higher number. We used 0 and 100 for value functions, but the use of 0 and 1 here will enable us to interpret what utilities actually represent. (If other values were used they could easily be transformed to a scale ranging from 0 to 1 without affecting the decision maker's preference between the courses of action.) Thus so far we have:

Monetary sum	Utility
$60 000	1.0
$30 000	Not yet known
$11 000	Not yet known
−$10 000	0

We now need to determine the business woman's utilities for the intermediate sums of money. There are several approaches which can be adopted to elicit utilities. The most commonly used methods involve offering the decision maker a series of choices between receiving given sums of money for certain or entering hypothetical lotteries. The decision maker's utility function is then inferred from the choices that are made. The method which we will demonstrate here is an example of the *probability-equivalence* approach (an alternative elicitation procedure will be discussed in a later section).

To obtain the business woman's utility for $30 000 using this approach we offer her a choice between receiving that sum for certain or entering a hypothetical lottery which will result in either the best outcome on the tree (i.e. a profit of $60 000) or the worst (i.e. a loss of $10 000) with specified probabilities. These probabilities are varied until the decision maker is indifferent between the certain money and the lottery. At this point, as we shall see, the utility can be calculated. A typical elicitation session might proceed as follows:

Question: Which of the following would you prefer?
  A $30 000 for certain; or
  B A lottery ticket which will give you a 70% chance of $60 000 and a 30% chance of −$10 000?

Answer: A 30% chance of losing $10 000 is too risky, I'll take the certain money.

We therefore need to make the lottery more attractive by increasing the probability of the best outcome.

Question: Which of the following would you prefer?
A $30 000 for certain; or
B A lottery ticket which will give you a 90% chance of $60 000 and a 10% chance of $-\$10\,000$?
Answer: I now stand such a good chance of winning the lottery that I think I'll buy the lottery ticket.

The point of indifference between the certain money and the lottery should therefore lie somewhere between a 70% chance of winning $60 000 (when the certain money was preferred) and a 90% chance (when the lottery ticket was preferred). Suppose that after trying several probabilities we pose the following question.

Question: Which of the following would you prefer?
A $30 000 for certain; or
B A lottery ticket which will give you an 85% chance of $60 000 and a 15% chance of $-\$10\,000$?
Answer: I am now indifferent between the certain money and the lottery ticket.

We are now in a position to calculate the utility of $30 000. Since the business woman is indifferent between options A and B the utility of $30 000 will be equal to the expected utility of the lottery. Thus:

$$u(\$30\,000) = 0.85u(\$60\,000) + 0.15u(-\$10\,000)$$

Since we have already allocated utilities of 1.0 and 0 to $60 000 and $-\$10\,000$, respectively, we have:

$$u(\$30\,000) = 0.85(1.0) + 0.15(0) = 0.85$$

Note that, once we have found the point of indifference, the utility of the certain money is simply equal to the probability of the best outcome in the lottery. Thus, if the decision maker had been indifferent between the options which we offered in the first question, her utility for $30 000 would have been 0.7.

We now need to determine the utility of $11 000. Suppose that after being asked a similar series of questions the business woman finally indicates that she would be indifferent between receiving $11 000 for certain and a lottery ticket offering a 60% chance of the best outcome ($60 000) and a 40% chance of the worst outcome (−$10 000). This implies that $u(\$11\,000) = 0.6$. We can now state the complete set of utilities and these are shown below:

Monetary sum	Utility
$60 000	1.0
$30 000	0.85
$11 000	0.60
−$10 000	0

These results are now applied to the decision tree by replacing the monetary values with their utilities (see Figure 4.3). By treating these utilities in the same way as the monetary values we are able to identify the course of action which leads to the highest expected utility.

Choosing the Luxuria Hotel gives an expected utility of:

$$0.6 \times 0.85 + 0.4 \times 0.6 = 0.75$$

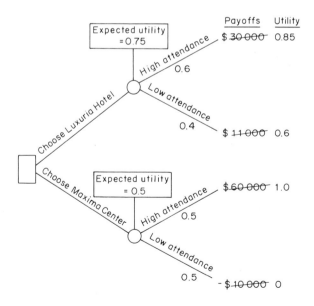

Figure 4.3 The conference organizer's decision tree with utilities

Choosing the Maxima Center gives an expected utility of:

$$0.5 \times 1.0 + 0.5 \times 0 = 0.5$$

Thus the business woman should choose the Luxuria Hotel as the venue for her exhibition. Clearly, the Maxima Center would be too risky.

It may be useful at this point to establish what expected utilities actually represent. Indeed, given that we have just applied the concept to a one-off decision, why do we use the term *expected* utility? To see what we have done, consider Figure 4.4(a). Here we have the business woman's decision tree with the original monetary sums replaced by the lotteries which she regarded as being equally attractive. For example, receiving $30 000 was considered to be equivalent to a lottery offering a 0.85 probability of $60 000 and a 0.15 probability of −$10 000. Obviously, receiving $60 000 is equivalent to a lottery ticket offering $60 000 for certain. You will see that every payoff in the tree is now expressed in terms of a probability of obtaining either the best outcome ($60 000) or the worst outcome (−$10 000).

Now, if the business woman chooses the Luxuria Hotel she will have a 0.6 probability of finishing with a profit which she perceives to be equivalent to a lottery ticket offering a 0.85 probability of $60 000 and a 0.15 probability of −$10 000. Similarly, she will have a 0.4 probability of a profit, which is equivalent to a lottery ticket offering a 0.6 probability of $60 000 and a 0.4 chance of −$10 000. Therefore the Luxuria Hotel offers her the equivalent of a $0.6 \times 0.85 + 0.4 \times 0.6 = 0.75$ probability of the best outcome (and a 0.25 probability of the worst outcome). Note that 0.75 is the expected utility of choosing the Luxuria Hotel.

Obviously, choosing the Maxima Center offers her the equivalent of only a 0.5 probability of the best outcome on the tree (and a 0.5 probability of the worst outcome). Thus, as shown in Figure 4.4(b), utility allows us to express the returns of all the courses of action in terms of simple lotteries all offering the same prizes, namely the best and worst outcomes, but with different probabilities. This makes the alternatives easy to compare. The probability of winning the best outcome in these lotteries is the expected utility. It therefore seems reasonable that we should select the option offering the highest expected utility.

Note that the use here of the term 'expected' utility is therefore somewhat misleading. It is used because the procedure for calculating expected utilities is arithmetically the same as that for calculating expected values in statistics. It does *not*, however, necessarily refer to an average result which would be obtained from a large number of repetitions of a course of action, nor does it mean a result or consequence which should be 'expected'. In decision theory, an 'expected utility' is only a 'certainty equivalent', that is, a single 'certain' figure that is equivalent in preference to the uncertain situations.

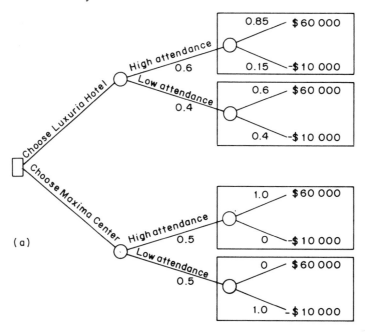

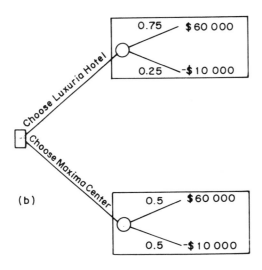

Figure 4.4 A demonstration of how expected utility reduces the decision to a simple choice between lotteries

INTERPRETING UTILITY FUNCTIONS

The business woman's utility function has been plotted on a graph in Figure 4.5. If we selected any two points on this curve and drew a straight line between them then it can be seen that the curve would always be above the line. Utility functions having this *concave* shape provide evidence of *risk aversion* (which is consistent with the business woman's avoidance of the riskiest option).

This is easily demonstrated. Consider Figure 4.6, which shows a utility function with a similar shape, and suppose that the decision maker, from whom this function has been elicited, has assets of $1000. He is then offered a gamble which will give him a 50% chance of doubling his money to $2000 and a 50% chance of losing it all, so that he finishes with $0. The expected monetary value of the gamble is $1000 (i.e. $0.5 \times \$2000 + 0.5 \times \$0$), so according to the EMV criterion he should be indifferent between keeping his money and gambling. However, when we apply the utility function to the decision we see that currently the decision maker has assets with a utility of 0.9. If he gambles he has a 50% chance of increasing his assets so that their utility would increase to 1.0 and a 50% chance of ending with assets with a utility of 0. Hence the expected utility of the gamble is $0.5 \times 1 + 0.5 \times 0$, which equals 0.5. Clearly, the certain money is more attractive than the risky option of gambling. In simple terms, even though the potential wins and losses are the same in monetary terms and even though he has the same chance of winning as he does of losing, the increase in utility which will occur if the decision maker wins the gamble is far less than the loss in utility he will

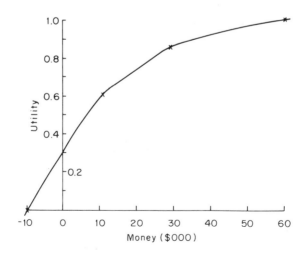

Figure 4.5 A utility function for the conference organizer

suffer if he loses. He therefore stands to lose much more than he stands to gain, so he will not be prepared to take the risk.

Figure 4.7 illustrates other typical utility functions. Figure 4.7(a) shows a utility function which indicates a risk-seeking attitude (or risk proneness). A person with a utility function like this would have accepted the gamble which we offered above. The linear utility function in Figure 4.7(b) demonstrates a risk-neutral attitude. If a person's utility function looks like this then the EMV criterion will represent their preferences. Finally, the utility function in Figure 4.7(c) indicates both a risk-seeking attitude and risk aversion. If the decision maker currently has assets of $y then he will be

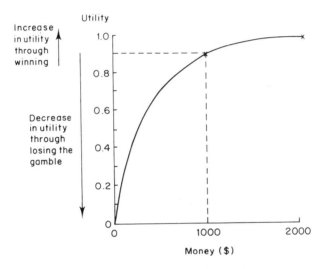

Figure 4.6 A utility function demonstrating risk aversion

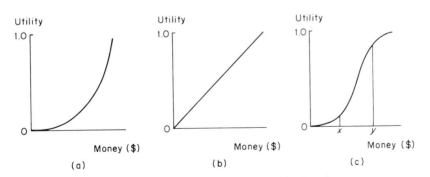

Figure 4.7 Interpreting the shape of a utility function. (a) A risk-seeking attitude; (b) risk neutrality, which means that the EMV criterion would represent the decision maker's preferences; (c) both a risk-seeking attitude and risk aversion

averse to taking a risk. The reverse is true if currently he has assets of only $x. It is important to note that individual's utility functions do not remain constant over time. They may vary from day to day, especially if the person's asset position changes. If you win a large sum of money tomorrow then you may be more willing to take a risk than you are today.

UTILITY FUNCTIONS FOR
NON-MONETARY ATTRIBUTES

Utility functions can be derived for attributes other than money. Consider the problem which is represented by the decision tree in Figure 4.8. This relates to a drug company which is hoping to develop a new product. If the company proceeds with its existing research methods it estimates that there is a 0.4 probability that the drug will take 6 years to develop and a 0.6 probability that development will take 4 years. However, recently a 'short-cut' method has been proposed which might lead to significant reductions in the development time, and the company, which has limited resources available for research, has to decide whether to take a risk and

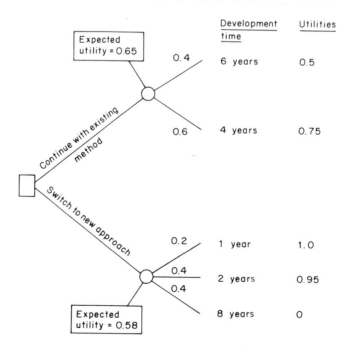

Figure 4.8 A decision tree for the drug company research department problem

switch completely to the proposed new method. The head of research estimates that, if the new approach is adopted, there is a 0.2 probability that development will take a year, a 0.4 probability that it will take 2 years and a 0.4 probability that the approach will not work and, because of the time wasted, it will take 8 years to develop the product.

Clearly, adopting the new approach is risky, so we need to derive utilities for the development times. The worst development time is 8 years, so $u(8$ years$)=0$ and the best time is 1 year, so $u(1$ year$)=1.0$. After being asked a series of questions, based on the variable probability method, the head of research is able to say that she is indifferent between a development time of 2 years and engaging in a lottery which will give her a 0.95 probability of a 1-year development and a 0.05 probability of an 8-year development time. Thus:

$$u(2 \text{ years})=0.95 \ u(1 \text{ year})+0.05 \ u(8 \text{ years})$$
$$=0.95(1.0)+0.05(0)=0.95$$

By a similar process we find that $u(4 \text{ years})=0.75$ and $u(6 \text{ years})=0.5$. The utilities are shown on the decision tree in Figure 4.8, where it can be seen that continuing with the existing method gives the highest expected utility. Note, however, that the two results are close, and a sensitivity analysis might reveal that minor changes in the probabilities or utilities would lead to the other alternative being selected. The utility function is shown in Figure 4.9. This has a concave shape indicating risk aversion.

It is also possible to derive utility functions for attributes which are not easily measured in numerical terms. For example, consider the choice of

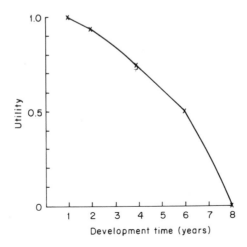

Figure 4.9 A utility function for product development time

design for a chemical plant. Design A may have a small probability of failure which may lead to pollution of the local environment. An alternative, design B, may also carry a small probability of failure which would not lead to pollution but would cause damage to some expensive equipment. If a decision maker ranks the possible outcomes from best to worst as: (i) no failure, (ii) equipment damage and (iii) pollution then, clearly, u(no failure) = 1 and u(pollution) = 0. The value of u(equipment damage) could then be determined by posing questions such as:

which would you prefer:
(1) A design which was certain at some stage to fail, causing equipment damage; or
(2) A design which had a 90% chance of not failing and a 10% chance of failing and causing pollution?

Once a point of indifference was established, u(equipment damage) could be derived.

Ronen *et al.*[2] describe a similar application in the electronics industry, where the decision relates to designs of electronic circuits for cardiac pacemakers. The designs carry a risk of particular malfunctions and the utilities relate to outcomes such as 'pacemaker not functioning at all', 'pacemaker working too fast', 'pacemaker working too slow' and 'pacemaker functioning OK'.

THE AXIOMS OF UTILITY

In the last few sections we have suggested that a rational decision maker should select the course of action which maximizes expected utility. This will be true if the decision maker's preferences conform to the following axioms:

Axiom 1: The complete ordering axiom
To satisfy this axiom the decision maker must be able to place all lotteries in order of preference. For example, if he is offered a choice between two lotteries, the decision maker must be able to say which he prefers or whether he is indifferent between them. (For the purposes of this discussion we will also regard a certain chance of winning a reward as a lottery.)

Axiom 2: The transitivity axiom
If the decision maker prefers lottery A to lottery B and lottery B to lottery C then, if he conforms to this axiom, he must also prefer lottery A to lottery C (i.e. his preferences must be transitive).

Axiom 3: The continuity axiom
Suppose that we offer the decision maker a choice between the two lotteries shown in Figure 4.10. This shows that lottery 1 offers a reward of B for certain while lottery 2 offers a reward of A, with probability p and a reward of C with probability $1-p$. Reward A is preferable to reward B, and B in turn is preferred to reward C. The continuity axiom states that there must be some value of p at which the decision maker will be indifferent between the two lotteries. We obviously assumed that this axiom applied when we elicited the conference organizer's utility for $30 000 earlier in the chapter.

Axiom 4: The substitution axiom
Suppose that a decision maker indicates that he is indifferent between the lotteries shown in Figure 4.11, where X, Y and Z are rewards and p is a probability. According to the substitution axiom, if reward X appears as a reward in another lottery it can always be substituted by lottery 2 because the decision maker regards X and lottery 2 as being equally preferable. For example, the conference organizer indicated that she was indifferent between the lotteries shown in Figure 4.12(a). If the substitution axiom applies, she will also be indifferent between lotteries 3 and 4, which are shown in Figure 4.12(b). Note that these lotteries are identical, except that in lottery 4 we have substituted lottery 2 for the $30 000. Lottery 4 offers a 0.6 chance of winning a ticket in another lottery and is therefore referred to as a *compound lottery*.

Axiom 5: Unequal probability axiom
Suppose that a decision maker prefers reward A to reward B. Then, according to this axiom, if he is offered two lotteries which only offer rewards A and B as possible outcomes he will prefer the lottery offering the highest probability of reward A. We used this axiom in our explanation of utility

Figure 4.10

Figure 4.11

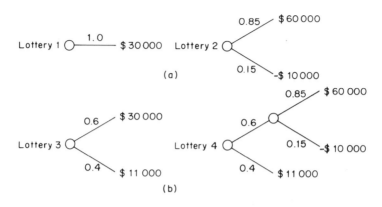

Figure 4.12

Figure 4.13

earlier, where we reduced the conference organizer's decision to a comparison of the two lotteries shown in Figure 4.13. Clearly, if the conference organizer's preferences conform to this axiom then she will prefer lottery 1.

Axiom 6: Compound lottery axiom

If this axiom applies then a decision maker will be indifferent between a compound lottery and simple lottery which offers the same rewards with the same probabilities. For example, suppose that the conference organizer is offered the compound lottery shown in Figure 4.14(a). Note that this lottery offers a 0.28 (i.e. 0.4×0.7) probability of $60 000 and a 0.72 (i.e. 0.4×0.3+0.6) probability of −$10 000. According to this axiom she will also be indifferent between the compound lottery and the simple lottery shown in Figure 4.14(b).

It can be shown (see, for example, French[3]) that if the decision maker accepts these six axioms than a utility function exists which represents his preferences. Moreover, if the decision maker behaves in a manner which is consistent with the axioms (i.e. rationally), then he will choose the course of action which has the highest expected utility. Of course, it may be possible to demonstrate that a particular decision maker does not act according to the axioms of utility theory. However, this does not necessarily imply that

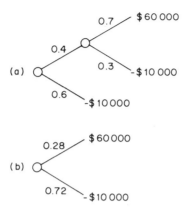

Figure 4.14

the theory is inappropriate in his case. All that is required is that he *wishes* to behave consistently according to the axioms. Applying decision analysis *helps* a decision maker to formulate preferences, assess uncertainty and make judgments in a coherent fashion. Thus coherence is the *result* of decision analysis, not a prerequisite.

MORE ON UTILITY ELICITATION

So far, we have only considered utility assessment based on the probability-equivalence approach. A disadvantage of this approach is that the decision maker may have difficulty in thinking in terms of probabilities like 0.90 or 0.95. Because of this, a number of alternative approaches have been developed (for example, Farquahar[4] reviews 24 different methods). Perhaps the most widely used of these is the *certainty-equivalence approach*, which, in its most common form, only requires the decision maker to think in terms of 50:50 gambles.

To illustrate the approach, let us suppose that we wish to elicit a decision maker's utility function for monetary values in the range $0–40 000 (so that $u($0)=0$ and $u($40 000)=1$). An elicitation session might proceed as follows:

Analyst: If I offered you a hypothetical lottery ticket which gave a 50% chance of $0 and a 50% chance of $40 000, how much would you be prepared to pay for it? Obviously, its expected monetary value is $20 000, but I want to know the minimum amount of money you would just be willing to pay for the ticket.
Decision maker: (after some thought) $10 000.

$$\text{Hence } u(\$10\,000) = 0.5\ u(\$0) + 0.5\ u(\$40\,000)$$
$$= 0.5(0) \quad + 0.5(1) \qquad\qquad = 0.5$$

The analyst would now use the $10 000 as the worst payoff in a new hypothetical lottery.

Analyst: If I now offered you a hypothetical lottery ticket which gave you a 50% chance of $40 000 and a 50% chance of $10 000 how much would you be prepared to pay for it?
Decision maker: About $18 000.

$$\text{Hence } u(\$18\,000) = 0.5\ u(\$10\,000) + 0.5\ u(\$40\,000)$$
$$= 0.5(0.5) \qquad + 0.5(1) \qquad\qquad = 0.75$$

The $10 000 is also used as the best payoff in a lottery which will also offer a chance of $0.

Analyst: What would you be prepared to pay for a ticket offering a 50% chance of $10 000 and a 50% chance of $0?
Decision maker: $3000.

$$\text{Thus } u(\$3000) = 0.5\ u(\$0) + 0.5\ u(\$10\,000)$$
$$= 0.5(0) \quad + 0.5(0.5) \qquad\qquad = 0.25$$

It can be seen that the effect of this procedure is to elicit the monetary values which have utilities of 0, 0.25, 0.5, 0.75 and 1. Thus we have:

Monetary value:	$0	$3000	$10 000	$18 000	$40 000
Utility	0	0.25	0.5	0.75	1.0

If we plotted this utility function on a graph it would be seen that the decision maker is risk averse for this range of monetary values. The curve could, of course, also be used to estimate the utilities of other sums of money.

While the certainty-equivalence method we have just demonstrated frees the decision maker from the need to think about awkward probabilities it is not without its dangers. You will have noted that the decision maker's first response ($10 000) was used by the analyst in subsequent lotteries, both as a best and worst outcome. This process is known as chaining, and the effect of this can be to propagate earlier judgmental errors.

The obvious question is, do these two approaches to utility elicitation produce consistent responses? Unfortunately, the evidence is that they do not. Indeed, utilities appear to be extremely sensitive to the elicitation method which is adopted. For example, Hershey et al.[5] identified a number of sources of inconsistency. Certainty-equivalence methods were found to yield greater risk seeking than probability-equivalence methods. The payoffs and probabilities used in the lotteries and, in particular, whether or not they included possible losses also led to different utility functions. Moreover, it was found that responses differed depending upon whether the choice offered involved risk being assumed or transferred away. For example, in the certainty-equivalence method we could either ask the decision maker how much he would be prepared to pay to *buy* the lottery ticket or, assuming that he already owns the ticket, how much he would accept to *sell* it. Research suggests that people tend to offer a lower price to buy the ticket than they would accept to sell it. There is thus a propensity to prefer the status quo, so that people are generally happier to retain a given risk than to take the same risk on (see also Thaler[6]). Finally, the context in which the questions were framed was found to have an effect on responses. For example, Hershey et al.[5] refer to an earlier experiment when the same choice was posed in different ways, the first involving an insurance decision and the second a gamble as shown below:

Insurance formulation
Situation A: You stand a one out of a thousand chance of losing $1000.
Situation B: You can buy insurance for $10 to protect you from this loss.
Gamble formulation
Situation A: You stand a one out of a thousand chance of losing $1000.
Situation B: You will lose $10 with certainty.

It was found that 81% of subjects preferred B in the insurance formulation, while only 56% preferred B in the gamble formulation.

Tversky and Kahneman[7] provide further evidence that the way in which the choice is framed affects the decision maker's response. They found that choices involving statements about gains tend to produce risk-averse responses, while those involving losses are often risk seeking. For example, in an experiment subjects were asked to choose a program to combat a disease which was otherwise expected to kill 600 people. One group was told that Program A would certainly save 200 lives while Program B offered a 1/3 probability of saving all 600 people and a 2/3 probability of saving nobody. Most subjects preferred A. A second group were offered the equivalent choice, but this time the statements referred to the number of deaths, rather than lives saved. They were therefore told that the first program would lead to 400 deaths while the second would offer a 1/3 probability of no deaths and a 2/3 probability of 600 deaths. Most subjects in this group preferred the second program, which clearly carries the higher risk. Further experimental evidence that different assessment methods lead to different utilities can be found in a paper by Johnson and Schkade.[8]

What are the implications of this research for utility assessment? First, it is clear that utility assessment requires effort and commitment from the decision maker. This suggests that, before the actual elicitation takes place, there should be a pre-analysis phase in which the importance of the task is explained to the decision maker so that he will feel motivated to think carefully about his responses to the questions posed.

Second, the fact that different elicitation methods are likely to generate different assessments means that the use of several methods is advisable. By posing questions in new ways the consistency of the original utilities can be checked and any inconsistencies between the assessments can be explored and reconciled.

Third, since the utility assessments appear to be very sensitive to both the values used and the context in which the questions are framed it is a good idea to phrase the actual utility questions in terms which are closely related to the values which appear in the original decision problem. For example, if there is no chance of losses being incurred in the original problem then the lotteries used in the utility elicitation should not involve the chances of incurring a loss. Similarly, if the decision problem involves only very high or low probabilities then the use of lotteries involving 50:50 chances should be avoided.

HOW USEFUL IS UTILITY IN PRACTICE?

We have seen that utility theory is designed to provide guidance on how to choose between alternative courses of action under conditions of uncertainty, but how useful is utility in practice? Is it really worth going to the trouble of asking the decision maker a series of potentially difficult questions about imaginary lotteries given that, as we have just seen, there are likely to be errors in the resulting assessments? We will summarize here arguments both for and against the application of utility and then present our own views at the end of the section.

First, let us restate that the *raison d'être* of utility is that it allows the attitude to risk of the decision maker to be taken into account in the decision model. Consider again the drug research problem which we discussed earlier. We might have approached this in three different ways. First, we could have simply taken the course of action which led to the shortest expected development time. These expected times would have been calculated as follows:

Expected development time of continuing with the existing method
$$=0.4\times6+0.6\times4 \qquad =4.8 \text{ years}$$

Expected development time of switching to new research approach
$$=0.2\times1+0.4\times2+0.4\times8=4.2 \text{ years}$$

The adoption of this criterion would therefore suggest that we should switch to the new research approach. However, this criterion ignores two factors. First, it assumes that each extra year of development time is perceived as being equally bad by the decision maker, whereas it is possible, for example, that an increase in time from 1 to 2 years is much less serious than an increase from 7 to 8 years. This factor could be captured by a *value function*. We could therefore have used one of the methods introduced in Chapter 2 to attach numbers on a scale from 0 to 100 to the different development times in order to represent the decision maker's relative preference for them. These *values* would then have replaced the actual development times in the calculations above and the course of action leading to the highest expected value could be selected. You will recall, however, from Chapter 2 that the derivation of a value function does not involve any considerations about probability, and it therefore will not capture the second omission from the above analysis, which is, of course, the attitude to risk of the decision maker. A utility function is therefore designed to allow *both* of these factors to be taken into account.

Despite this, there are a number of arguments against the use of utility. Perhaps the most persuasive relates to the problems of measuring utility. As Tocher[9] has argued, the elicitation of utilities takes the decision maker away from the real world of the decision to a world of hypothetical lotteries. Because these lotteries are only imaginary, the decision maker's judgments about the relative attractiveness of the lotteries may not reflect what he would really do. It is easy to say that you are prepared to accept a 10% risk of losing $10 000 in a hypothetical lottery, but would you take the risk if you were really facing this decision? Others (e.g. von Winterfeldt and Edwards[10]) argue that if utilities can only be measured approximately then it may not always be worth taking the trouble to assess them since a value function, which is more easily assessed, would offer a good enough approximation. Indeed, even Howard Raiffa,[11] a leading proponent of the utility approach, argues:

> Many analysts assume that a value scoring system—designed for tradeoffs under certainty—can also be used for probabilistic choice (using expected values). Such an assumption is wrong theoretically, but as I become more experienced I gain more tolerance for these analytical simplifications. This is, I believe, a relatively benign mistake in practice.

Another criticism of utility relates to what is known as Allais's paradox. To illustrate this, suppose that you were offered the choice of options A and B as shown in Figure 4.15(a). Which would you choose? Experiments suggest that most people would choose A (e.g. see Slovic and Tversky[12]). After all, $1 million for certain is extremely attractive while option B offers only a small probability of $5 million and a chance of receiving $0.

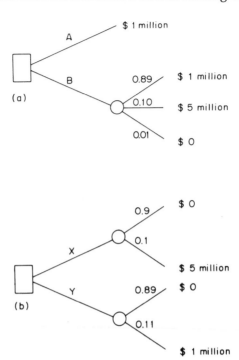

Figure 4.15 Allais's paradox

Now consider the two options X and Y which are shown in Figure 4.15(b). Which of these would you choose? The most popular choice in experiments is X. With both X and Y, the chances of winning are almost the same, so it would seem to make sense to go for the option offering the biggest prize.

However, if you did choose options A and X your judgments are in conflict with utility theory, as we will now show. If we let $u(\$5m)=1$ and $u(\$0)=0$, then selecting option A suggests that:

$u(\$1m)$ is greater than $0.89\ u(\$1m)+0.1\ u(\$5m)+0.01\ u(\$0m)$
i.e. $u(\$1m)$ exceeds $0.89\ u(\$1m)+0.1$ which implies:
$u(\$1m)$ *exceeds* $0.1/0.11$

However, choosing X implies that:

$0.9\ u(\$0)+0.1\ u(\$5m)$ exceeds $0.89\ u(\$0)+0.11\ u(\$1m)$
i.e. 0.1 exceeds $0.11\ u(\$1m)$
so that: $u(\$1m)$ is *less than* $0.1/0.11$.

This paradox has stimulated much debate[13] since it was put forward in 1953. However, we should emphasize that utility theory does not attempt to describe the way in which people make decisions like those posed above. It is intended as a normative theory, which indicates what a rational decision maker should do *if* he accepts the axioms of the theory. The fact that people make inconsistent judgments does not by itself invalidate the theory. Nevertheless, it seems sensible to take a relaxed view of the problem. Remember that utility theory is designed as simply an aid to decision making, and if a decision maker wants to ignore its indications then that is his prerogative.

Having summarized some of the main arguments, what are our views on the practical usefulness of utility? First, we have doubts about the practice adopted by some analysts of applying utility to decisions where risk and uncertainty are not central to the decision maker's concerns. Introducing questions about lotteries and probabilities to these sorts of problems seems to us to be unnecessary. In these circumstances the problem of trading off conflicting objectives is likely to be the main concern, and we would therefore recommend the approach of Chapter 2. In important problems which do involve a high level of uncertainty and risk we do feel that utility has a valuable role to play as long as the decision maker is familiar with the concept of probability, and has the time and patience to devote the necessary effort and thought to the questions required by the elicitation procedure. In these circumstances the derivation of utilities may lead to valuable insights into the decision problem. In view of the problems associated with utility assessment, we should not regard the utilities as perfect measures and automatically follow the course of action they prescribe. Instead, it is more sensible to think of the utility function as a useful tool for gaining a greater understanding of the problem.

If the decision maker does not have the characteristics outlined above or only requires rough guidance on a problem then it may not be worth eliciting utilities. Given the errors which are likely to occur in utility assessment, the derivation of values (as opposed to utilities) and the identification of the course of action yielding the highest expected value may offer a robust enough approach. (Indeed, there is evidence that linear utility functions are extremely robust approximations.) Sensitivity analysis would, of course, reveal just how precise the judgments needed to be.

In the final section of this chapter we extend the application of utility to problems involving more than one attribute. We should point out that multi-attribute utility analysis can be rather complex and the number of people applying it is not large. In the light of this, and the points made in our discussion above, we have decided to give only an introduction to this area so that a general appreciation can be gained of the type of judgments required.

MULTI-ATTRIBUTE UTILITY

So far in this chapter we have focused on decision problems which involve uncertainty and only one attribute. We next examine how problems involving uncertainty and multiple attributes can be handled. In essence, the problem of deriving a multi-attribute utility function is analogous to that of deriving a multi-attribute value function, which we discussed in Chapter 2. Again, the 'divide and conquer' philosophy applies. As we argued before, large multi-faceted problems are often difficult to grasp in their entirety. By dividing the problem into small parts and allowing the decision maker to focus on each small part separately we aim to simplify his judgmental task. Thus if certain conditions apply, we can derive a single attribute utility function for each attribute using the methods of earlier sections and then combine these to obtain a multi-attribute utility function. A number of methods have been proposed for performing this analysis, but the approach we will discuss is associated with Keeney and Raiffa.[14] This approach has been applied to decision problems ranging from the expansion of Mexico City Airport (de Neufville and Keeney[15]) to the selection of sites for nuclear power plants (Kirkwood[16]).

The Decanal Engineering Corporation

To illustrate the approach let us consider the following problem which involves just two attributes. The Decanal Engineering Corporation has recently signed a contract to carry out a major overhaul of a company's equipment. Ideally, the customer would like the overhaul to be completed in 12 weeks and, if Decanal meet the target or do not exceed it by a significant amount of time, they are likely to gain a substantial amount of goodwill from the customer and an enhanced reputation throughout the industry. However, to increase the chances of meeting the target, Decanal would have to hire extra labor and operate some 24-hour working, which would increase their costs. Thus the company has two conflicting objectives: (1) minimize the time that the project overruns the target date and (2) minimize the cost of the project.

For simplicity, we will assume that Decanal's project manager has two options: (1) work normally or (2) hire extra labor and work 24-hour shifts. His estimates of the probabilities that the project will overrun the target date by a certain number of weeks are shown on the decision tree in Figure 4.16. The costs of the project for the two options and for different project durations are also shown on the tree. (Note that, once a given option is chosen, the longer the project takes to complete, the greater will be the costs because labor, equipment, etc. will be employed on the project for a longer period.)

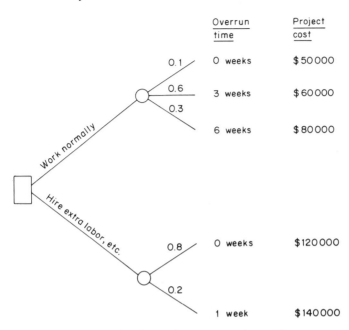

Figure 4.16   A decision tree for the project manager's problem

To analyze this problem we need to derive a multi-attribute utility function which will enable the project manager to compare the two options. This process is simplified if certain assumptions can be made. The most important of these is that of mutual utility independence.

Mutual Utility Independence

Suppose that the project manager is indifferent between the following alternatives:

A: A project which will certainly overrun by 2 weeks and which will certainly cost $50 000; and
B: A gamble which will give him a 50% chance of a project which overruns by 0 weeks (i.e it meets the target) and which will cost $50 000 and a 50% chance of a project which will overrun by 6 weeks and cost $50 000.

These alternatives are shown in Figure 4.17(a) (note that all the costs are the same).
Suppose that we now offer the project manager the same two options, but with the project costs increased to $140 000, as shown in Figure 4.17(b). If the project manager is still indifferent between the options then clearly his preference between the overrun times is unaffected by the change in

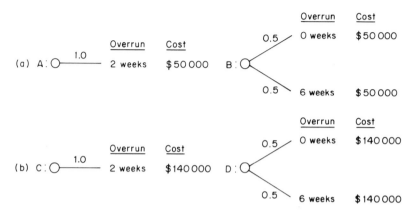

Figure 4.17

costs. If this is the case for all possible costs then overrun time is said to be *utility independent* of project cost. Putting this in more general terms: attribute A is utility independent of attribute B if the decision maker's preferences between gambles involving different levels of A, but the same level of B, do not depend on the level of attribute B.

It can be seen that utility independence is analogous to preference independence, which we discussed in Chapter 2, except that we are now considering problems which involve uncertainty. If project cost is also utility independent of overrun time (this will not automatically be the case) then we can say that overrun time and project cost are *mutually utility independent*.

The great advantage of mutual utility independence, if it exists, is that it enables the decision maker to concentrate initially on deriving utility function for one attribute at a time without the need to worry about the other attributes. If this independence does not exist then the analysis can be extremely complex (see Keeney and Raiffa[14]), but in very many practical situations it is usually possible to define the attributes in such a way that they do have the required independence.

Deriving the Multi-attribute Utility Function

Assuming that mutual utility independence does exist, we now derive the multi-attribute utility function as follows.

Stage 1: Derive single-attribute utility functions for overrun time and project cost.

Stage 2: Combine the single-attribute functions to obtain a multi-attribute utility function so that we can compare the alternative courses of action in terms of their performance over both attributes.

Stage 3: Perform consistency checks, to see if the multi-attribute utility function really does represent the decision maker's preferences, and sensitivity analysis to examine the effect of changes in the figures supplied by the decision maker.

Stage 1

First we need to derive a utility function for project overrun. Using the approach which we discussed earlier in the context of single-attribute utility, we give the best overrun (0 weeks) a utility of 1.0 and the worst (6 weeks) a utility of 0. We then attempt to find the utility of the intermediate values, starting with an overrun of 3 weeks. After being asked a series of questions, the project manager indicates that he is indifferent between:

A: A project which will certainly overrun by 3 weeks; and
B: A gamble offering a 60% chance of a project with 0 weeks overrun and a 40% chance of a 6-week overrun.

This implies that u(3-weeks overrun)=0.6. By a similar process, the manager indicates that u(1-week overrun)=0.9. The resulting utility function is shown in Figure 4.18(a).

We then repeat the elicitation process to obtain a utility function for project cost. The function obtained from the manager is shown in Figure 4.18(b). Table 4.3 summarizes the utilities which have been elicited for overrun and cost.

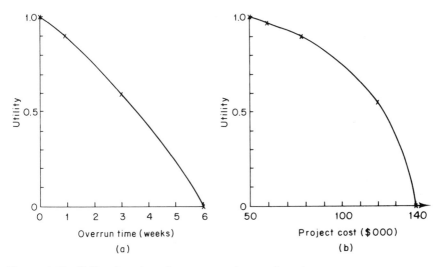

Figure 4.18 Utility functions for overrun time and project cost

Table 4.3 The project manager's utilities for overrun and cost

No. of weeks project overruns target	Utility	Cost of project ($)	Utility
0	1.0	50 000	1.0
1	0.9	60 000	0.96
3	0.6	80 000	0.9
6	0.0	120 000	0.55
		140 000	0.0

Stage 2

We now need to combine these utility functions to obtain the multi-attribute utility function. If the two attributes are mutually utility independent then it can be shown that the multi-attribute utility function will have the following form:

$$u(x_1,x_2) = k_1 u(x_1) + k_2 u(x_2) + k_3 u(x_1)u(x_2)$$

where

$x_1 =$ the level of attribute 1,

$x_2 =$ the level of attribute 2,

$u(x_1,x_2) =$ the multi-attribute utility if attribute 1 has a level x_1 and attribute 2 has a level x_2,

$u(x_1) =$ the single-attribute utility if attribute 1 has a level x_1,

$u(x_2) =$ the single-attribute utility if attribute 2 has a level x_2

and k_1, k_2, and k_3 are numbers which are used to 'weight' the single-attribute utilities.

In stage 1 we derived $u(x_1)$ and $u(x_2)$, so we now need to find the values of k_1, k_2 and k_3. We note that k_1 is the weight attached to the utility for overrun time. In order to find its value we offer the project manager a choice between the following alternatives:

A: A project where overrun is certain to be at its best level (i.e. 0 weeks), but where the cost is certain to be at its worst level (i.e. $140 000); or

B: A lottery which offers a probability of k_1 that both cost and overrun will be at their best levels (i.e. 0 weeks and $50 000) and a $1-k_1$ probability that they will both be at their worst levels (i.e. 6 weeks and $140 000, respectively).

These options are shown in Figure 4.19. Note that because we are finding k_1 it is attribute 1 (i.e. overrun) which appears at its best level in the certain outcome.

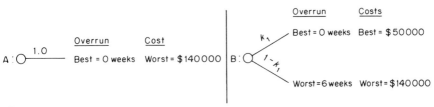

Figure 4.19

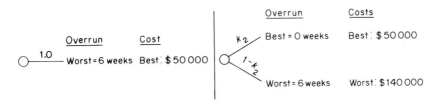

Figure 4.20

The decision maker is now asked what value the probability k_1 must have to make him indifferent between the certain outcome and the lottery. After some thought, he indicates that this probability is 0.8, so $k_1=0.8$. This suggests that the 'swing' from the worst to the best overrun time is seen by the project manager to be significant relative to project cost. If he hardly cared whether the overrun was 0 or 6 weeks, it would have taken only a small value of k_1 to have made him indifferent to a gamble where overrun time might turn out to be at its worst level.

To obtain k_2, the weight for project cost, we offer the project manager a similar pair of options. However, in the certain outcome project cost is now at its best level and the other attribute at its worst level. The probability of the best outcome in the lottery is now k_2. These two options are shown in Figure 4.20.

We now ask the project manager what value k_2 would need to be to make him indifferent between the two options. He judges this probability to be 0.6, so $k_2=0.6$. The fact that k_2 is less than k_1 suggests that the project manager sees the swing from the worst to the best cost as being less significant than the swing from the worst to the best overrun time. Having been offered a project which is certain to incur the lowest cost, he requires a smaller probability to tempt him to the lottery, where he might gain a project where overrun is also at its best level but where there is also a risk of a project with costs at their worst level.

Finally, we need to find k_3. This is a simple calculation and it can be shown that:

$$k_1+k_2+k_3=1, \text{ so } k_3=1-k_1-k_2$$

Thus, in our case $k_3 = 1 - 0.8 - 0.6 = -0.4$. The project manager's multi-attribute utility function is therefore:

$$u(x_1, x_2) = 0.8 \ u(x_1) + 0.6 \ u(x_2) - 0.4 \ u(x_1)u(x_2)$$

We can now use the multi-attribute utility function to determine the utilities of the different outcomes in the decision tree. For example, to find the utility of a project which overruns by 3 weeks and costs \$60 000 we proceed as follows. From the single-attribute functions we know that $u(3$ weeks overrun$) = 0.6$ and $u(\$60\,000$ cost$) = 0.96$. Therefore:

$$u(3 \text{ weeks overrun, } \$60\,000 \text{ cost})$$
$$= 0.8 \ u(3 \text{ weeks overrun}) + 0.6 \ u(\$60\,000 \text{ cost})$$
$$- 0.4 \ u(3 \text{ weeks overrun}) \ u(\$60\,000 \text{ cost})$$

$$= 0.8(0.6) + 0.6(0.96) - 0.4(0.6)(0.96) = 0.8256$$

Figure 4.21 shows the decision tree again with the multi-attribute utilities replacing the original attribute values. By multiplying the probabilities of the outcomes by their utilities we obtain the expected utility

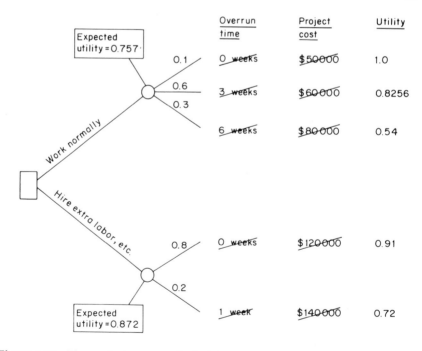

Figure 4.21 The project manager's decision tree with utilities

of each option. The results shown on the tree indicate that the project manager should hire the extra labor and operate 24-hour working, as this yields the highest expected utility.

Stage 3

It is important that we should check that the results of the analysis have faithfully represented the project manager's preferences. This can involve tracking back through the analysis and explaining why one option has performed well and another has performed badly. If the decision maker does not feel that the explanations are consistent with his preferences then the analysis may need to be repeated. In fact, it is likely that several iterations will be necessary before a consistent representation is achieved and, as the decision maker gains a greater understanding of his problem, he may wish to revise his earlier responses.

Another way of checking consistency is to offer the decision maker a new set of lotteries and to ask him to rank them in order of preference. For example, we could offer the project manager the three lotteries shown in Figure 4.22. The expected utilities of these lotteries are A: 0.726, B: 0.888 and C: 0.620, so if he is consistent then he should rank them in the order B, A, C. We should also carry out sensitivity analysis on the probabilities and utilities by, for example, examining the effect of changes in the values of k_1 and k_2.

Interpreting Multi-attribute Utilities

In the analysis above we derived an expected utility of 0.872 for the 'hire extra labor . . . ' option, but what does this figure actually represent? We demonstrated earlier that we could use the concept of utility to convert a decision problem to a simple choice between lotteries which the decision maker regarded as being equivalent to the original outcomes. Each of these lotteries would result in either the best or worst possible outcome, but with different probabilities. The same is true for multi-attribute utility. This time the lotteries will result in either the best outcome on both attributes (i.e. the best/best outcome) or the worst possible outcome on both attributes (i.e.

Figure 4.22

worst/worst). Thus the expected utility of 0.872 for the 'hire extra labor . . . ' option implies that the decision maker regards this option as being equivalent to a lottery offering a 0.872 chance of the best/best outcome (and a complementary probability of the worst/worst outcome). It therefore seems reasonable that he should prefer this option to the 'work normally' alternative, which is regarded as being equivalent to a lottery offering only a 0.757 chance of the best/best outcome.

Further Points on Multi-attribute Utility

The principles which we applied to the two-attribute problem above can be extended to any number of attributes (see, for example, Bunn[17] who discusses a problem involving four attributes), though the form of the multi-attribute utility function becomes more complex as the number of attributes increases. Models have also been developed which can handle situations where mutual utility independence does not exist (see Keeney and Raiffa[14]), but the complexities of these models have meant that they have proved to be of little practical value. In any case, as we mentioned earlier, if mutual utility independence does not exist it is likely that by redefining the attributes a new set can be found which do exhibit the required independence (we discussed the analogous problem in Chapter 2 when looking at multi-attribute value functions).

The approach to multi-attribute utility which we discussed above clearly requires a major commitment of time and effort from the decision maker and, since the method lacks the 'transparency' of the SMART procedure, which we met in Chapter 2, a non-mathematical person may be suspicious of its results. In all models a balance has to be struck between the accuracy with which the model represents the real problem and the effort required to formulate the model. If a problem is of major importance, and if the decision maker is happy to make the necessary judgments, then what Watson and Buede[18] refer to as the 'deep soul searching' engendered by Keeney and Raiffa's approach may lead to valuable insights into the decision problem. In other circumstances, where the decision maker only requires outline guidance from the model, a less sophisticated approach based, for example, on values rather than utilities may suffice. Sensitivity analysis will provide useful guidance on the robustness of any approximations which are used.

SUMMARY

In this chapter we have considered a number of methods which enable a decision maker to make rational decisions when the outcomes of courses of action are not known for certain. The approach based on expected monetary value was the simplest, but if the decision maker does not have a neutral attitude to risk, then the adoption of this criterion may lead to the most-preferred course of action not being chosen. We therefore introduced the concept of expected utility to show how the decision maker's attitude to risk can be incorporated into the decision model. Finally, we showed how the application of utility can be extended to decision problems involving more than one attribute.

EXERCISES

(1) An entertainment company is organizing a pop concert in London. The company has to decide how much it should spend on publicizing the event and three options have been identified:

Option 1: Advertise only in the music press;
Option 2: As option 1 but also advertise in the national press;
Option 3: As options 1 and 2 but also advertise on commercial radio.

For simplicity, the demand for tickets is categorized as low, medium or high. The payoff table below shows how the profit which the company will earn for each option depends on the level of demand.

				Profits ($000s)
Option		Demand		
	Low	Medium	High	
1	− 20	20	100	
2	− 60	− 20	60	
3	− 100	− 60	20	

It is estimated that if option 1 is adopted the probabilities of low, medium and high demand are 0.4, 0.5 and 0.1, respectively. For option 2 the respective probabilities are 0.1, 0.3 and 0.6 while for option 3 they are 0.05, 0.15 and 0.8. Determine the option which will lead to the highest expected profit. Would you have have any reservations about recommending this option to the company?

(2) A speculator is considering the purchase of a commodity which he reckons has a 60% chance of increasing in value over the next month. If he purchases the commodity and it does increase in value the speculator will make a profit of about $200 000, otherwise he will lose $60 000.

(a) Assuming that the expected monetary value criterion is applicable, determine whether the speculator should purchase the commodity.

(b) Perform a sensitivity analysis on the speculator's estimate of the probability of a price increase and interpret your result.

(c) What reservations would you have about applying the expected monetary value criterion in this context?

(3) A team of scientists is due to spend six months in Antarctica carrying out research. One major piece of equipment they will be taking is subject to breakdowns caused by the sudden failure of a particular component. Because a failed component cannot be repaired the team intend to carry a stock of spare units of the component, but it will cost them roughly $3000 for each spare unit they take with them. However, if the equipment breaks down and a spare is not available a new unit will have to be specially flown in and the team will incur a total cost of $4000 for each unit that is delivered in this way. An engineer who will be travelling with the team has estimated that the number of spares that will be required during the six months follows the probability distribution shown below:

No. of spares required	0	1	2	3
Probability	0.2	0.3	0.4	0.1

Determine the number of spares that the team should carry if their objective is to minimize expected costs.

(4) You are a contestant on a television game show and you have won £5000 so far. You are now offered a choice: either you can keep the money and leave or you can continue into the next round, where you have a 70% chance of increasing your winnings to $10 000 and a 30% chance of losing the $5000 and finishing the game with nothing.

(a) Which option would you choose?

(b) How does your choice compare with that which would be prescribed by the expected monetary value criterion?

(5) A building contractor is submitting an estimate to a potential customer for carrying out some construction work at the customer's premises. The builder reckons that if he offers to carry out the work for $150 000 there is a 0.2 probability that the customer will agree to the price, a 0.5 probability that a price of $120 000 would eventually be agreed and a 0.3 probability that the customer will simply refuse the offer and give the work to another builder. If the builder offers to carry out the work for $100 000 he reckons that there is a 0.3 probability that the customer will accept this price, a 0.6 probability that the customer will bargain so that a price of $80 000 will eventually be agreed and a 0.1 probability that the customer will refuse the offer and take the work elsewhere.

(a) Determine which price the builder should quote in order to maximize the expected payment he receives from the customer.

(b) Suppose that, after some questioning, the builder is able to make the following statements:

'I am indifferent between receiving $120 000 for certain or entering a lottery that will give me a 0.9 probability of $150 000 and a 0.1 probability of winning $0.'

'I am indifferent between receiving $100 000 for certain or entering a lottery that will give me a 0.85 probability of winning $150 000 and a 0.15 probability of winning $0.'

'I am indifferent between receiving $80 000 for certain or entering a lottery that will give me a 0.75 probability of winning $150 000 and a 0.25 probability of winning $0.'

   (i) Sketch the builder's utility function and comment on what it shows.
   (ii) In the light of the above statements which price should the builder now quote to the customer and why?

(6) (a) Use the following questions to assess your own utility function for money values between $0 and $5000. You should assume that all sums of money referred to will be received immediately.
   (i) You are offered either a sum of money for certain or a lottery ticket that will give you a 50% chance of winning $5000 and a 50% chance of winning $0. Write down below the certain sum of money which would make you indifferent between whether you received it or the lottery ticket.
      $ . . . . (we will now refer to this sum of money as X)
      The utility of X is 0.5.
   (ii) You are now offered a lottery ticket which offers you a 50% chance of $ . . . . (enter X here) and a 50% chance of $0. Alternatively, you will receive a sum of money for certain. Write down below the certain sum of money which would make you indifferent between whether you received it or the lottery ticket.
      $ . . . .
      The utility of this sum of money is 0.25.
   (iii) Finally, you are offered a sum of money for certain or a lottery ticket which will give you a 50% chance of $5000 and a 50% chance of $ . . . . (enter X here). Write down below the certain sum of money which would make you indifferent between whether you received it or the lottery ticket.
      $ . . . .
      The utility of this sum of money is 0.75.
(b) Plot your utility function and discuss what it reveals.

(c) Discuss the strengths and limitations of the assessment procedure which was used in (a).

(7) A company is planning to re-equip one of its major production plants and one of two types of machine, the Zeta and the Precision II, is to be purchased. The prices of the two machines are very similar so the choice of machine is to be based on two factors: running costs and reliability. It is agreed that these two factors can be represented by the variables: average weekly operating costs and number of breakdowns in the first year of operation. The company's production manager estimates that the following probability distributions apply to the two machines. It can be assumed that the probability distributions for operating costs and number of breakdowns are independent.

Zeta

Average weekly operating costs ($)	Prob.	No of breakdowns	Prob.
20 000	0.6	0	0.15
30 000	0.4	1	0.85

Precision II

Average weekly operating costs ($)	Prob.	No. of breakdowns	Prob.
15 000	0.5	0	0.2
35 000	0.5	1	0.7
		2	0.1

Details of the manager's utility functions for operating costs and number of breakdowns are shown below:

Average weekly operating costs ($)	Utility	No. of breakdowns	Utility
15 000	1.0	0	1.0
20 000	0.8	1	0.9
30 000	0.3	2	0
35 000	0		

(a) The production manager's responses to questions reveal that, for him, the two attributes are mutually utility independent. Explain what this means.

(b) The production manager also indicates that for him $k_1=0.7$ (where attribute 1=operating costs) and $k_2=0.5$. Discuss how these values could have been determined.

(c) Which machine has the highest expected utility for the production manager?

(8) The managers of the Lightning Cycle Company are hoping to develop a new bicycle braking system. Two alternative systems have been proposed and, although the mechanics of the two systems are similar, one design will use mainly plastic components while the other will use mainly metal ones. Ideally, the design chosen would be the lightest and the most durable but, because some of the technology involved is new, there is some uncertainty about what the characteristics of the resulting product would be.

The leader of Lightning's research and development team has estimated that if the plastic design is developed there is a 60% chance that the resulting system would add 130 grams to a bicycle's weight and would have a guaranteed lifetime of one year. He also reckons that there is a 40% chance that a product with a 2-year lifetime could be developed, but this would weigh 180 grams.

Alternatively, if the metal design was developed the team leader estimates that there is a 70% chance that a product with a 2-year guaranteed life and weighing 250 grams could be developed. However, he estimates that there is a 30% chance that the resulting product would have a guaranteed lifetime of 3 years and would weigh 290 grams.

It was established that, for the team leader, weight and guaranteed lifetime were mutually utility independent. The following utilities were then elicited from him:

Weight (grams)	Utility	Guaranteed lifetime (years)	Utility
130	1.0	3	1.0
180	0.9	2	0.6
250	0.6	1	0
290	0		

After further questioning the team leader indicated that he would be indifferent between the following alternatives:

A: A product which was certain to weigh 130 grams, but which had a guaranteed lifetime of only one year; or

B: A gamble which offered a 0.7 probability of a product with a weight of 130 grams and a guaranteed lifetime of 3 years and a 0.3 probability of a product with a weight of 290 grams and a guaranteed lifetime of one year. $\therefore K_1 = 0.7$

Finally, the team leader said that he would be indifferent between alternatives C and D below:

C: A product which was certain to weigh 290 grams, but which had a guaranteed lifetime of 3 years;

D: A gamble which offered a 0.9 probability of a product with a weight of 130 grams and a guaranteed lifetime of 3 years and a 0.1 probability of a product with a weight of 290 grams and a guaranteed lifetime of one year. $k_2 = 0.9$

(a) What do the team leader's responses indicate about his attitude to risk and the relative weight which he attaches to the two attributes of the proposed design?

(b) Which design should the team leader choose, given the above responses?

(c) What further analysis should be conducted before a firm recommendation can be made to the team leader?

REFERENCES

1. Von Neumann, J. and Morgenstern, O. (1944) *The Theory of Games and Economic Behavior*, Princeton University Press.
2. Ronen, B., Pliskin, J. S. and Feldman, S. (1984) Balancing the Failure Modes in the Electronic Circuit of a Cardiac Pacemaker: a Decision Analysis, *Journal of the Operational Research Society*, **35**, No. 5, 379–387.
3. French, S. (1988) *Decision Theory: An Introduction to the Mathematics of Rationality*, Ellis Horwood, Chichester.
4. Farquahar, P. H. (1984). Utility Assessment Methods, *Management Science*, **30**, No. 11, 1283–1300.
5. Hershey, J. C., Kunreuther, H. C. and Schoemaker, P. J. H. (1982) Sources of Bias in Assessment Procedures for Utility Functions, *Management Science*, **28**, No. 8, 936–954.
6. Thaler, R. H. (1983) Illusions and Mirages in public policy, in H. R. Arkes and K. R. Hammond (eds) *Judgment and Decision Making*, Cambridge University Press, Cambridge.
7. Tversky, A. and Kahneman, D. (1981) The Framing of Decisions and the Psychology of Choice, *Science*, **211**, 30 January, 453–458.
8. Johnson, E. J. and Schkade, D. A. (1989) Bias in Utility Assessments: Further Evidence and Explanations, *Management Science*, **35**, No. 4, 406–424.
9. Tocher, K. D. (1977) Planning Systems, *Philosophical Transactions of the Royal Society of London*, **A287**, 425–441.
10. Von Winterfeldt, D. and Edwards, W. (1986) *Decision Analysis and Behavioral Research*, Cambridge University Press, Cambridge.
11. Raiffa, H. (1982) *The Art and Science of Negotiation*, Harvard University Press, Cambridge, Mass.
12. Slovic, P. and Tversky A. (1974) Who Accepts Savage's Axiom? *Behavioral Science*, **19**, 368–373.
13. Indeed, a whole debate has been organized around paradoxical choice and the psychological acceptability of the axioms of expected utility theory and its variants. See, for example, Stigum, B. P. and Wenstop, F. (eds) (1983) *Foundations of Risk and Utility Theory with Applications*, Reidel, Dordrecht.
14. Keeney, R. L. and Raiffa, H. (1976) *Decisions with Multiple Objectives: Preferences and Value Tradeoffs*, Wiley, New York.

15. De Neufville, R. and Keeney, R. L. (1972) Systems Evaluation through Decision Analysis: Mexico City Airport, *Journal of Systems Engineering*, **3**, No. 1, 34–50.

16. Kirkwood, C. W. (1982) A Case History of Nuclear Power Plant Site Selection, *Journal of the Operational Research Society*, **33**, 353–363.

17. Bunn, D. W. (1982) *Analysis for Optimal Decisions*, Wiley, Chichester.

18. Watson, S. R. and Buede, D. M. (1987) *Decision Synthesis*, Cambridge University Press, Cambridge.

Chapter 5

Decision Trees and Influence Diagrams

INTRODUCTION

When they are first encountered, some decision problems appear to be overwhelmingly complex. Any attempt at clear thinking can be frustrated by the large number of interrelated elements which are associated with the problem so that, at best, the unaided decision maker can have only a hazy perception of the issues involved. In these circumstances, decision trees and influence diagrams can be extremely useful in helping people to gain an understanding of the structure of the problems which confront them.

We have already introduced some very simple decision trees in Chapter 4, but here we will extend the idea to show how multi-stage problems can be modeled. Decision problems are multi-stage in character when the choice of a given option may result in circumstances which will require yet another decision to be made. For example, a company may face an immediate decision relating to the manufacturing capacity which should be provided for a new product. Later, when the product has been on the market for several years, it may have to decide whether to expand or reduce the capacity. This later decision will have to be borne in mind when the initial decision is being made, since the costs of converting from one capacity level to another may vary. A decision to invest now in a very small manufacturing plant might lead to high costs in the future if a major expansion is undertaken. This means that the decisions made at the different points in time are interconnected.

As we will see, decision trees can serve a number of purposes when complex multi-stage problems are encountered. They can help a decision maker to develop a clear view of the structure of a problem and make it easier to determine the possible scenarios which can result if a particular course of action is chosen. This can lead to creative thinking and the generation of options which were not previously being considered. Decision trees can also help a decision maker to judge the nature of the information which needs to be gathered in order to tackle a problem and, because they are generally easy to understand, they can be an excellent medium for communicating one person's perception of a problem to other individuals.

The process of constructing a decision tree is usually iterative, with many changes being made to the original structure as the decision maker's understanding of the problem develops. Because the intention is to help the decision maker to think about the problem, very large and complex trees, which are designed to represent every possible scenario which can occur, can be counterproductive. Decision trees are models, and as such are simplifications of the real problem. The simplification is the very strength of the modeling process because it fosters the understanding and insight which would be obscured by detail and complexity.

Influence diagrams offer an alternative way of structuring a complex decision problem and some analysts find that people relate to them much more easily. Indeed Howard[1] has called them: 'The greatest advance I have seen in the communication, elicitation and detailed representation of human knowledge . . . the best tool I know of for crossing the bridge from the original opaque situation in the person's mind to a clear and crisp decision basis.' As we shall show later, influence diagrams can be converted to decision trees and we will therefore regard them in this chapter as a method for eliciting decision trees. However, some computer programs now exist which use complex algorithms to enable the influence diagram to be used not just as an initial elicitation tool but as a means for identifying the best sequence of decisions.

CONSTRUCTING A DECISION TREE

You may recall from earlier chapters that two symbols are used in decision trees. A square is used to represent a decision node and, because each branch emanating from this node represents an option, the decision maker can choose which branch to follow. A circle, on the other hand, is used to represent a chance node. The branches which stem from this sort of node represent the possible outcomes of a given course of action and the branch which is followed will be determined, not by the decision maker, but by circumstances which lie beyond his or her control. The branches emanating from a circle are therefore labeled with probabilities which represent the decision maker's estimate of the probability that a particular branch will be followed. Obviously, it is not sensible to attach probabilities to the branches which stem from a square.

The following example will be used to demonstrate how a decision tree can be used in the analysis of a multi-stage problem. An engineer who works for a company which produces equipment for the food-processing industry has been asked to consider the development of a new type of processor and to make a recommendation to the company's board. Two alternative power sources could be used for the processor, namely gas and electricity, but for technical reasons each power source would require a fundamentally different design. Resource constraints mean that the company will only be able to pursue one of the designs, and because the processor would be more advanced than others which have been developed it is by no means certain that either design would be a success. The engineer estimates that there is a 75% chance that the electricity-powered design would be successful and only a 60% chance that the gas-powered design would be a success.

Figure 5.1 shows an initial decision tree for the problem with estimated payoffs in millions of dollars. After considering this tree the engineer realizes that if either design failed then the company would still consider modifying the design, though this would involve more investment and would still not guarantee success. He estimates that the probability that the electrical design could be successfully modified is only 30%, though the gas design would have an 80% chance of being modified successfully. This leads to the new tree which is shown in Figure 5.2. Note that the decision problem is now perceived to have two stages. At stage one a decision has to be made between the designs or not developing the problem at all. At stage two a decision *may* have to be made on whether the design should be modified.

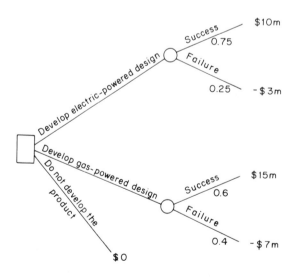

Figure 5.1 An initial decision tree for the food-processor problem

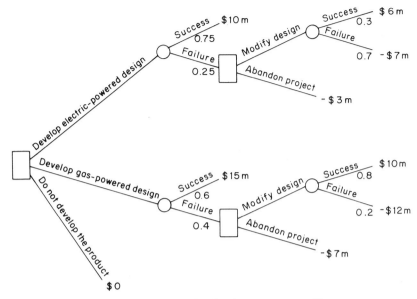

Figure 5.2   A new decision tree for the food-processor problem

After some reflection, the engineer agrees that this tree is a satisfactory representation of the options facing the company. Other alternatives such as switching to the development of a gas-powered design if the electrical design is not successful are not considered to be feasible, given the resources available to the company.

DETERMINING THE OPTIMAL POLICY

It can be seen that our decision tree consists of a set of *policies*. A policy is a plan of action stating which option is to be chosen at each decision node that might be reached under that policy. For example, one policy would be: choose the electrical design; if it fails, modify the design. Another policy would be: choose the electrical design; if it fails, abandon the project.

We will now show how the decision tree can be used to identify the optimal policy. For simplicity, we will assume that the engineer considers that monetary return is the only attribute which is relevant to the decision, and we will also assume that, because the company is involved in a large number of projects, it is neutral to the risk involved in this development and therefore the expected monetary value (EMV) criterion is appropriate. Considerations of the timing of the cash flows and the relative preference for receiving cash flows at different points in time will also be excluded from our analysis (this issue is dealt with in Chapter 6).

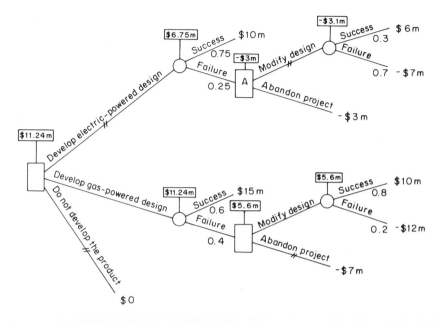

Figure 5.3 Rolling back the decision tree

The technique for determining the optimal policy in a decision tree is known as the *rollback method*. To apply this method, we analyze the tree from right to left by considering the later decisions first. The process is illustrated in Figure 5.3. Thus if the company chose the electrical design and it failed (i.e. if the decision node labeled with an A was reached), what would be the best course of action? Modifying the design would lead to an expected return of $(0.3 \times \$6m) + (0.7 \times -\$7m)$, which equals $-\$3.1m$. Since this is worse than the $-\$3m$ payoff that would be achieved if the design was abandoned, abandoning the design would be the best course of action. Two bars are therefore placed over the inferior branch and the 'winning' payoff is moved back to the decision node where it is now treated as a payoff for the 'failure' branch. This means that the expected payoff of the electrical design is $(0.75 \times \$10m) + (0.25 \times -\$3m)$, which equals $\$6.75m$.

The same analysis is applied to the section of the tree that represents the gas-powered design. It can be seen that if this design fails the best option is to modify it. Hence the expected payoff of the gas design is $\$11.24m$. This exceeds the expected payoff of the electrical design and the $\$0$ payoff of not proceeding with the development. Two bars are therefore placed over the branches representing these options and the $\$11.24m$ is moved back to the initial decision node. The optimum policy is therefore to develop the gas-powered design and, if it fails, to modify the design.

It can be seen that the rollback method allows a complex decision problem to be analyzed as a series of smaller decision problems. We should, of course, now apply sensitivity analysis to the probabilities and payoffs using the method we introduced in the previous chapter. For brevity, this analysis will not be carried out here. It should also be pointed out that the decision tree suggests the best policy based on the information which is available at the time it is constructed. By the time the engineer knows whether or not the gas-powered design is successful his perception of the problem may have changed and he would then, of course, be advised to review the decision. For example, if the design fails, the knowledge he has gained in attempting to develop the equipment may lead him to conclude that modification would be unlikely to succeed and he might then recommend abandonment of the project.

Note also that the planning period which the tree represents is arbitrary. Even if a successful gas design is developed this surely will not be the end of the story, since this choice of design is bound to have ramifications in the future. For example, any money earned from the design may well be re-invested in research and development for future products and the developments of these products may or may not be successful, and so on. Moreover, if the company chooses to develop its knowledge of gas, rather than electric, technology, this may restrict its options in the longterm. However, any attempt to formulate a tree which represents every possible consequence and decision which may arise over a period stretching into the distant future would clearly lead to a model which was so complex that it would be intractable. Judgment is therefore needed to determine where the tree should end.

Clearly, the calculations involved in analyzing a large decision tree can be rather tedious. Because of this, a number of computer packages have been developed which will display and analyze decision trees and allow them to be easily modified. ARBORIST, developed by Texas Instruments Inc. (Austin, Texas), and SUPERTREE, developed by SDG Decision Systems (Menlo Park, California), are two such packages. In addition, Jones[2] has shown how spreadsheet packages can be used in the construction and analysis of decision trees. However, the procedures involved can be rather time consuming.

DECISION TREES AND UTILITY

In the previous section we made the assumption that the decision maker was neutral to risk. Let us now suppose that the engineer is concerned that his career prospects will be blighted if the development of the processor leads to a great loss of money for the company. He is therefore risk averse, and

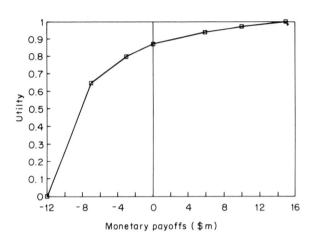

Figure 5.4 The engineer's utility function

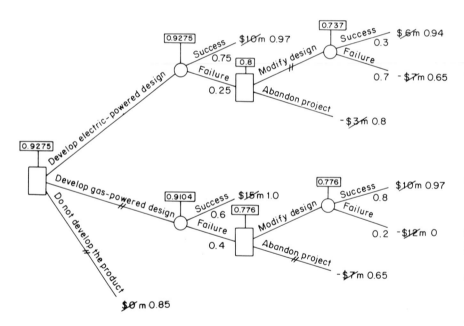

Figure 5.5 Applying the rollback method to a decision tree involving utilities

his utility function for the monetary sums involved in this problem is shown in Figure 5.4.

The procedure for analyzing the tree when utilities are involved is exactly the same as that which we used for the EMV criterion. Figure 5.5 shows the decision tree, with the utilities replacing the monetary values. After applying the rollback method it can be seen that now the optimum policy is to develop the electric-powered design and, if it fails, to abandon the project. Note, however, that the closeness of the expected utilities suggests that sensitivity analysis should be applied to the tree before a firm decision is made.

If the engineer had wished to include other attributes besides money in his decision model then multi-attribute utilities would have appeared at the ends of the tree. However, the rollback procedure would still have been applied in the same way. This would also be the case if the payoffs on the tree had been represented as net present values (see Chapter 6).

DECISION TREES INVOLVING CONTINUOUS PROBABILITY DISTRIBUTIONS

In the decision problem we considered above there were only two possible outcomes for each course of action, namely success and failure. However, in some problems the number of possible outcomes may be very large or even infinite. Consider, for example, the possible percentage market share a company might achieve after an advertising campaign or the possible levels of cost which may result from the development of a new product. Variables like these could be represented by continuous probability distributions, but how can we incorporate such distributions into our decision tree format? One obvious solution is to use a discrete probability distribution as an approximation. For example, we might approximate a market share distribution with just three outcomes: high, medium and low. A number of methods for making this sort of approximation have been suggested, and we will discuss the *Extended Pearson–Tukey (EP–T) approximation* here. This was proposed by Keefer and Bodily,[3] who found it to be a very good approximation to a wide range of continuous distributions. The method is based on earlier work by Pearson and Tukey,[4] and requires three estimates to be made by the decision maker:

(i) The value in the distribution which has a 95% chance of being exceeded. This value is allocated a probability of 0.185.
(ii) The value in the distribution which has a 50% chance of being exceeded. This value is allocated a probability of 0.63.
(iii) The value in the distribution which has only a 5% chance of being exceeded. This value is also allocated a probability of 0.185.

To illustrate the method, let us suppose that a marketing manager has to decide whether to launch a new product and wishes to represent on a decision tree the possible sales levels which will be achieved in the first year

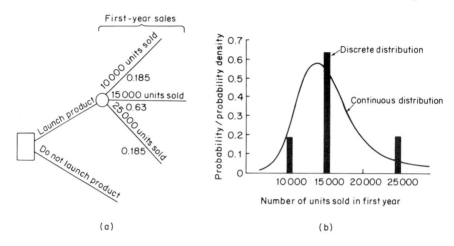

(a)   (b)

Figure 5.6 The extended Pearson–Tukey (EP–T) approximation method

if the product is launched. To apply the EP–T approximation to the sales probability distribution we would need to obtain the three estimates from the decision maker. Suppose that she estimates that there is a 95% chance that first-year sales will exceed 10 000 units, a 50% chance that they will exceed 15 000 units and a 5% chance they will exceed 25 000 units. The resulting decision tree is shown in Figure 5.6(a), while Figure 5.6(b) illustrates how the discrete distribution has been used to approximate the continuous distribution.

Of course, in many decision trees the probability distributions will be dependent. For example, in our product launch example one might expect second-year sales to be related to the sales which were achieved in the first year. In this case, questions like the following would need to be asked to obtain the distribution for second-year sales: 'Given that first-year sales were around 25 000 units, what level of sales in the second year would have a 50% chance of being exceeded?'

As Keefer and Bodily point out, the EP–T approximation does have its limitations. It would be inappropriate to use it where the continuous probability distribution has more than one peak (or mode) and the approximation would probably not be a good one if the shape of the continuous distribution was very asymmetric. Moreover, in some decision problems a subsequent decision depends upon the achievement of a particular level of a variable. For example, in our product launch problem the manager may decide to discontinue the product after the first year if sales do not reach 12 000 units. In this case, clearly attention should be focused on the probability of this critical sales level being reached, rather than on the three points used in the EP–T approximation. Nevertheless, in general, there are clear advantages in using this approximation. Above all, it is simple and each distribution requires only three estimates to be made which has the obvious effect of reducing the decision maker's judgmental task.

PRACTICAL APPLICATIONS OF DECISION TREES

A large number of applications of decision trees have been published over the years, and we give below a summary of a few of these applications to show the variety of contexts where the method has been successfully used.

Ulvila[5] used the technique to help the US postal service to decide on whether to continue with the nine-digit zip code for business users. The analysis was designed to compare the monetary returns which might result from the use of various types of automatic sorting equipment either with or without the code. The EP–T approximation was used to represent probability distributions of the extent to which the code would be used, and the savings which would result from the various options in the tree. The author reported that the approach helped the decision makers 'to think creatively about the problem and to generate options'.

Cohan *et al.*[6] used decision trees to analyze decisions relating to the use of fire in forest management. For example, forest fires are deliberately used to control pests and diseases, but a decision to start a controlled fire at a particular time has to be made in the light of uncertainty about weather conditions and the actual success of the fire. Nevertheless, if a decision is made to postpone a fire this itself will result in costs caused by the delay in meeting the foresters' objectives. The authors reported that the use of decision analysis helped forest managers to understand the implications and relative importance of the key uncertainties involved in the decisions. Moreover, it provided 'clear and concise documentation of the rationale underlying important decisions'. They argued that 'this can be invaluable in stimulating discussion and communication within a forest management organization'.

Bell[7] used a decision tree to help New England Electric (a public utility) to decide on the price they should bid at an auction for the salvage rights of a ship which had run aground off the coast of Florida. If the bid succeeded a decision would have to be made on whether to install self-loading equipment on the ship, while if it failed a decision would have to be made on what type of ship to purchase instead. The intention was to use the ship to carry coal and other goods on the open market, and there were uncertainties relating to factors such as whether a given bid price would win and the level of revenue which would be received from renting out the ship.

Other published applications include a decision relating to the choice between two sites for drilling an oil well (Hosseini[8]), the problem faced by the management of a coal-fired power plant in evaluating and selecting particulate emission control equipment (Madden *et al.*[9]), management–union bargaining (Winter[10]) and the problem of reorganizing the subsidiaries of a company, all of which required new premises (Hertz and Thomas[11]).

As we have seen, decision trees are the major analytical structures underlying application of decision analysis to problems involving uncertainty. In the examples that we have used so far in this book we have either given a decision tree representation or used a case example where the individual pieces in the jigsaw were sufficient and necessary to complete the case analysis for subsequent computations. Real-life decision problems may, at first pass, contain *pieces from many different jigsaws*. The trick is to know which pieces are missing (and so need to be obtained) or which are either redundant or not relevant to the problem analysis in hand.

ASSESSMENT OF DECISION STRUCTURE

Consider the following 'real-life' decision problem that we would like you to attempt to represent in the form of a decision tree.

Imagine that you are a businessman and you are considering making electronic calculators. Your factory can be equipped to manufacture them and you recognize that other companies have profited from producing them. However, equipping the factory for production will be very expensive and you have seen the price of calculators dropping steadily. What should you do?

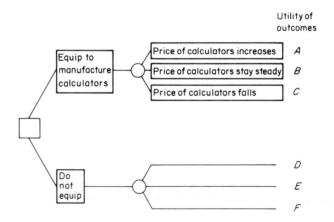

Figure 5.7 One decision-analytic representation of the calculator problem

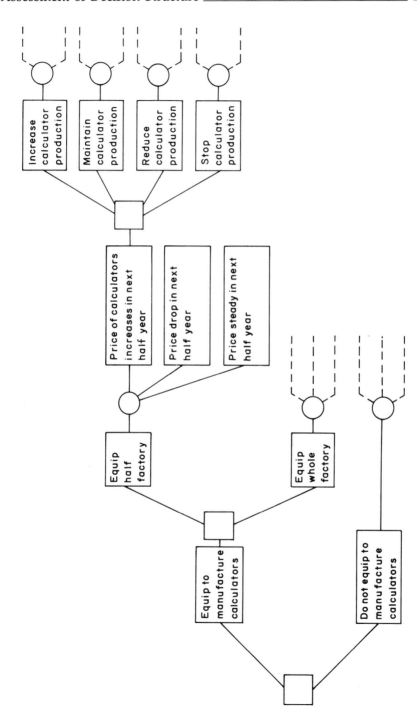

Figure 5.8 Toward the correct decision-analytic representation of the calculator problem?

Well, what is the correct decision-analytic representation? Figure 5.7 presents one representation which may or may not match yours. Figure 5.8 is a more elaborate and perhaps more realistic representation of the problem.

Do you agree? Actually, as you have probably guessed, there is no *obviously* right or wrong representation of any problem that is in any way related to real life. Although Expected Utility may be an optimal decision principle there is no normative technique for eliciting the *structure* of the decision problem from the decision maker. It is really a matter of the decision analyst's judgment as to whether the elicited tree is a fair representation of the decision maker's decision problem. Once a structure is agreed then the computation of expected utility is fairly straightforward. Structuring is therefore a major problem in decision analysis, for if the structuring is wrong then it is a

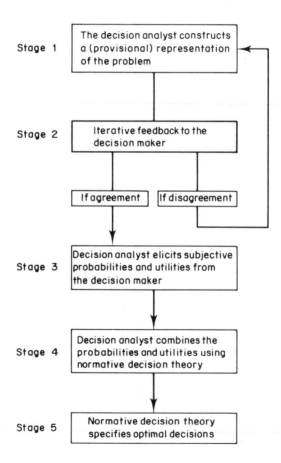

Figure 5.9 Phases of a decision analysis

necessary consequence that assessments of utilities and probabilities may be inappropriate and the expected utility computations may be invalid.

Figure 5.9 presents a description of the typical phases in a decision analysis of a problem that the decision maker wishes to resolve with the help of the practitioner of decision analysis—the decision analyst.

Stages 1 and 2 of the decision process are iterative, the structure of the decision problem emerges from discussions between the decision maker and the analyst. Once a structure for the decision representation has been agreed and probabilities and utilities are elicited (stage 3) the expected utility for the various acts under consideration can be computed (stage 4) and the act which has the maximal expected utility is chosen (stage 5).

What determines the decision analyst's provisional representation of the decision problem? Generally, it will be based upon past experience with similar classes of decision problems and, to a significant extent, intuition. To quote Von Winterfeldt:[12]

> Often the analyst decides on a specific structure and later finds it unmanageable . . . knowing about the recursive nature of the structuring process, it is good decision analysis practice to spend much effort on structuring and to keep an open mind about possible revisions.

However, problem representation is an art rather than a science, as Fischhoff[13] notes:

> Regarding the validation of particular assessment techniques we know . . . next to nothing about eliciting the structure of problems from decision-makers.

Keeney[14] has fewer reservations:

> Often the complex problems are so involved that their structure is not well understood. A simple decision tree emphasizing the problem structure which illustrates the main alternatives, uncertainties, and consequences, can usually be drawn up in a day. Not only does this often help in defining the problem, but it promotes client and colleague confidence that perhaps decision analysis can help. It has often been my experience that sketching out a simple decision tree with a client in an hour can lead to big advances in the eventual solution to a problem.

Many decision makers report that they feel the process of problem representation is perhaps more important than the subsequent computations. Humphreys[15] has labeled the latter the 'direct value' of decision analysis and the former the 'indirect value'. Decision analysis provides the decision maker with a

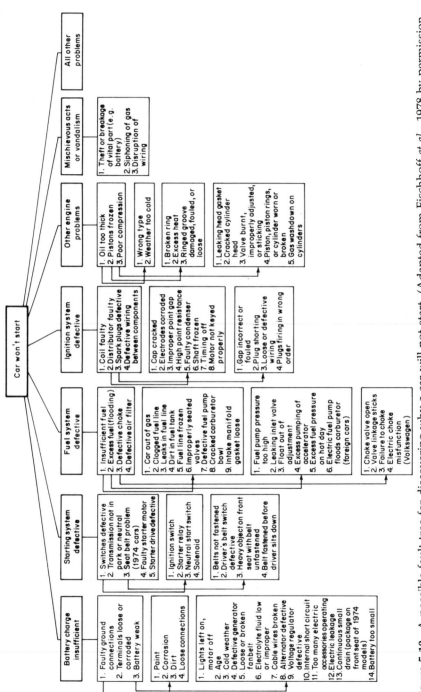

Figure 5.10 A possible fault tree for discovering why a car will not start. (Adapted from Fischhoff *et al.*, 1978 by permission of the authors)

convincing rationale for choice, improves communication and permits direct and separate comparisons of different people's conceptions of the structure of the problem, and of the assessment of decomposed elements within their structures, thereby raising consciousness about the root of any conflict.

However, some studies have illustrated that the decision makers' estimates, judgment and choices are affected by the way knowledge is elicited.

This research has direct relevance for the decision analyst's attempts at structuring. In one study, Fischhoff et al.[16] investigated estimation of failure probabilities in decision problem representations called fault trees. These fault trees are essentially similar to decision trees, with the exception that events rather than acts and events are represented. Figure 5.10 gives a fault tree representation for the event 'a car fails to start'. This is the full version of the fault tree that Fischhoff produced from the use of several car-repair reference texts.

In several experiments Fischhoff presented various 'full' and 'pruned' fault trees to members of the public. For example, three of the first six sub-events in Figure 5.10 would be omitted from the presentation to be implicitly included under the seventh sub-event 'all other problems'. Fischhoff asked:

For every 100 times that a trip is delayed due to a 'starting failure' estimate, on average, how many of the delays are caused by the 7(4) factors?

Fischhoff found that the amount of probability placed on 'all other problems' did not increase significantly when it contained three of the other main sub-events. In a subsequent experiment the importance of 'all other problems' was emphasized:

In particular we would like you to consider its [the fault tree's] completeness. That is, what proportion of the possible reasons for a car not starting are left out, to be included in the category, 'all other problems'?

However, focusing subjects' attention on what was missing only partially improved their awareness. Fischhoff labeled this insensitivity to the incompleteness of the fault tree 'out of sight, out of mind'. The finding was confirmed with technical experts and garage mechanics. Neither self-rated degree of knowledge nor actual garage experience has any significant association with subjects' ability to detect what was missing from the fault tree.

Another finding from the study was that the perceived importance of a particular sub-event or branch of the fault tree was increased by presenting it in pieces (i.e. as two separate branches). The implications of this result are far reaching. Decision trees constructed early in the analyst/decision maker interaction may be incomplete representations of the decision problem facing the decision maker.

ELICITING DECISION TREE REPRESENTATIONS

What methods have been developed to help elicit decision tree representations from decision makers? One major method, much favoured by some decision analysts, is that of *influence diagrams*,[17] which are designed to summarize the dependencies that are seen to exist among events and acts within a decision. Such dependencies may be mediated by the flow of time, as we saw in our examples of decision trees. As we shall see, a close relationship exists between influence diagrams and the more familiar decision trees. Indeed, given certain conditions, influence diagrams can be converted to trees. The advantage of starting with influence diagrams is that their

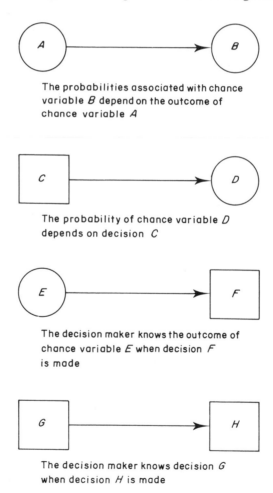

The probabilities associated with chance
variable *B* depend on the outcome of
chance variable *A*

The probability of chance variable *D*
depends on decision *C*

The decision maker knows the outcome of
chance variable *E* when decision *F*
is made

The decision maker knows decision *G*
when decision *H* is made

Figure 5.11 Definitions used in influence diagrams

graphic representation is more appealing to the intuition of decision makers who may be unfamiliar with decision technologies. In addition, influence diagrams are more easily revised and altered as the decision maker iterates with the decision analyst. Decision trees, because of their strict temporal ordering of acts and events, need completely respecifying when additional acts and events are inserted into preliminary representations. We shall illustrate the applicability of influence diagrams through a worked example. First, however, we will present the basic concepts and representations underlying the approach.

Figure 5.11 presents the key concepts. As with the decision tree, event nodes are represented by circles and decision nodes by squares. Arrowed lines between nodes indicate the influence of one node to another. For example, an arrow pointing to an event node indicates that either the likelihood of events (contained in the node) are influenced by a prior decision or on the occurrence (or not) of prior events. Alternatively, an arrow pointing to a decision node indicates that *either* the decision is influenced by a prior decision *or* on the occurrence (or not) of prior events. The

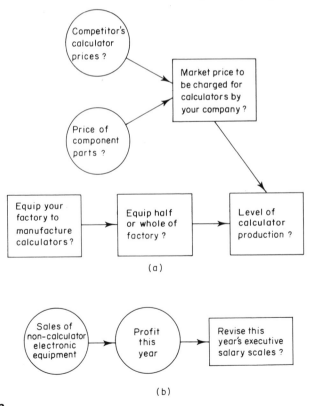

(a)

(b)

Figure 5.12

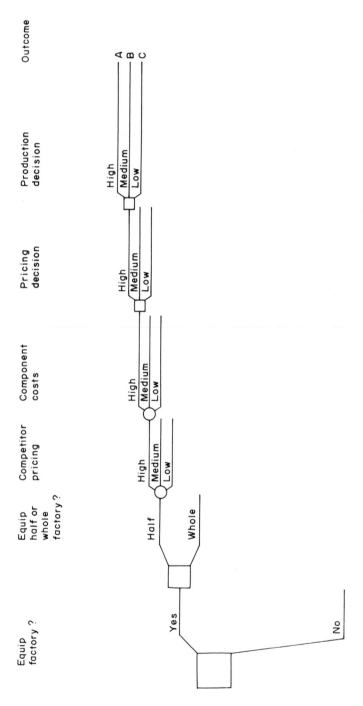

Figure 5.13

whole set of interconnected decisions and events is called an influence diagram.

Figure 5.12(a) gives one examplar influence diagram for the calculator production problem and Figure 5.13 gives the decision tree representation of this influence diagram. Two conditions must be met in order for an influence diagram to be represented as a decision tree. First, the diagram must contain no circles of influence arrows (loops). A loop would show that a node (variable) both influences and is influenced by another node. Such a relationship could not be represented by the left-to-right ordering of influences represented by a decision tree. Second, since decision trees are conventionally used to aid a *single* decision maker who is knowledgeable about *all* temporally prior acts and events, then those nodes that have *direct* influence (shown by direct arrows) on a particular decision must be a subset of the influences on subsequent decisions. If such an ordering is not represented in an influence diagram (for example, the combination of Figures 5.12(a) and 5.12(b)) then at least two decision trees need to be developed to represent the diagram.

Obviously, a decision tree representation must preserve the ordering represented by the arrows in an influence diagram and the tree must not have an event node as a predecessor of a decision node for which it is not directly linked by an arrow in the influence diagram. If the tree did, it would imply that the decision depends on the event node which, from the influence diagram, is not the case.

One step-by-step procedure for turning an influence diagram into a decision tree is as follows:

(1) Identify a node with no arrows pointing into it (since there can be no loops at least one node will be such).
(2) If there is a choice between a decision node and an event node, choose the decision node.
(3) Place the node at the beginning of the tree and 'remove' the node from the influence diagram.
(4) For the now-reduced diagram, choose another node with no arrows pointing into it. If there is a choice a decision node should be chosen.
(5) Place this node next in the tree and 'remove' it from the influence diagram.
(6) Repeat the above procedure until all the nodes have been removed from the influence diagram.

For practice, try this procedure on the content of Figure 5.12(a). You should achieve the decision tree represented in Figure 5.13. To complete the tree, the possible choices at each decision node and the possible events at each event node must now be specified. Finally, subjective probabilities must be assessed for the events and utilities must be assigned to the end points in the decision tree.

Very complex decision trees can be represented as one-page influence diagrams. However, the use of influence diagrams to construct decision trees where subsequent events and acts depend on the initial decision (i.e. where the resulting decision tree is asymmetric) is more problematic. In these instances, the influence diagram approach to decision tree structuring can be used as a guide only.

From our overview of influence diagrams you will have seen that such diagrams aid subsequent structuring of decision trees. They allow the easy insertion of additional acts and events as the decision maker talks through the decision problem with the decision analyst. (See stages 1 and 2 of Figure 5.9.) *By themselves, influence diagrams do not aid in the creation of decision options or in the recognition of event possibilities.* Such creation and recognition activities perhaps may be best thought of as creative behavior. As we have seen, Fischhoff *et al.* found that people seem to suffer from 'out of sight, out of mind' bias when evaluating the completeness of decision tree-type representations of knowledge.

In other words, individual decision makers may be inappropriately content with decision problem representations that are created early in the decision maker/analyst interaction. One major way to combat this tendency is to subject initial problem representations to outside critique by other people with a knowledge of the decision problem domain. Such critiques are readily available in the context of decision conferencing, where those individuals with a stake in a key decision interact with the aid of a decision analyst who acts to facilitate a decision. We will deal with this approach in detail in Chapter 11, where we will focus on the advantages and disadvantages of group decision making.

SUMMARY

In this chapter we have illustrated the construction of decision trees and rollback method for identifying the optimal policy. We described an approximation method for dealing with continuous probability distributions within decision trees and summarized some practical applications of decision trees within decision analysis. Finally, we analyzed the process of generating decision tree representation of decision problems and advocated the influence diagram as a key technique to facilitate decision structuring.

EXERCISES

(1) A company has to decide whether to invest money in the development of a microbiological product. The company's research director has estimated

that there is a 60% chance that a successful development could be achieved in two years. However, if the product had not been successfully developed at the end of this period, the company would abandon the project, which would lead to a loss in present value* terms of $3 million.

In the event of a successful development a decision would have to be made on the scale of production. The returns generated would depend on the level of sales which could be achieved over the period of the product's life. For simplicity, these have been categorized as either high or low. If the company opted for large-volume production and high sales were achieved then net returns with a present value of $6 million would be obtained. However, large-scale production followed by low sales would lead to net returns with a present value of only $1 million.

On the other hand, if the company decided to invest only in small-scale production facilities then high sales would generate net returns with a present value of $4 million and low sales would generate net returns with a present value of $2 million. The company's marketing manager estimates that there is a 75% chance that high sales could be achieved.

(a) Construct a decision tree to represent the company's decision problem.
(b) Assuming that the company's objective is to maximize its expected returns, determine the policy that it should adopt.
(c) There is some debate in the company about the probability that was estimated by the research director. Assuming that all other elements of the problem remain the same, determine how low this probability would have to be before the option of not developing the product should be chosen.
(d) Before the final decision is made the company is taken over by a new owner, who has the utilities shown below for the sums of money involved in the decision. (The owner has no interest in other attributes which may be associated with the decision such as developing a prestige product or maintaining employment.) What implications does this have for the policy that you identified in (b) and why?

Present value of net returns	New owner's utility
−$3m	0
$0	0.6
$1m	0.75
$2m	0.85
$4m	0.95
$6m	1.0

---

*Present value is designed to take the company's time preference for money into account. The concept is explained in Chapter 6.

(2) A large machine in a factory has broken down and the company that owns the factory will incur costs of $3200 for each day the machine is out of action. The factory's engineer has three immediate options:

Option 1
He can return the machine to the supplier who has agreed to collect, repair and return it free of charge, but not to compensate the company for any losses they might incur while the repair is being carried out. The supplier will *not* agree to repair the machine if any other person has previously attempted to repair it. If the machine is returned, the supplier will guarantee to return it in working order in 10 days' time.

Option 2
He can call in a specialist local engineering company. They will charge $20 000 to carry out the repair and they estimate that there is a 30% chance that they will be able to return the machine to working order in 2 days. There is, however, a 70% chance that repairs will take 4 days.

Option 3
He can attempt to carry out the repair work himself, and he estimates that there is a 50% chance that he could mend the machine in 5 days. However, if at the end of 5 days the attempted repair has not been successful he will have to decide whether to call in the local engineering company or to make a second attempt at repair by investigating a different part of the mechanism. This would take 2 further days, and he estimates that there is a 25% chance that this second attempt would be successful. If he fails at the second attempt, he will have no alternative other than to call in the local engineering company. It can be assumed that the probability distribution for the local engineering company's repair time will be unaffected by any work which the factory engineer has carried out.

Assuming that the engineer's objective is to minimize expected costs, what course(s) of action should he take?

(3) Westward Magazine Publishers are thinking of launching a new fashion magazine for women in the under-25 age group. Their original plans were to launch in April of next year, but information has been received that a rival publisher is planning a similar magazine. Westward now have to decide whether to bring their launch forward to January of next year, though this would cost an additional $500 000. If the launch is brought forward it is estimated that the chances of launching before the rival are about 80%. However, if the launch is not brought forward it is thought that there is only a 30% chance of launching before the rival.

For simplicity, the management of Westward have assumed that the circulation of the magazine throughout its life will be either high or low. If Westward launch before the rival, it is thought that there is a 75% chance of a high circulation. However, if the rival launches first, this probability is estimated to be only 50%.

If the rival does launch first then Westward could try to boost sales by increasing their level of advertising. This would cost an extra $200 000, but it is thought that it would increase the probability of a high circulation to 70%. This increased advertising expenditure would not be considered if Westward's magazine was launched first.

Westward's accountants have estimated that a high circulation would generate a gross profit over the magazine's lifetime of $4 million. A low circulation would bring a gross profit of about $1 million. It is important to note, however, that these gross profits do *not* take into account additional expenditure caused by bringing the launch forward or by increased advertising.

(a) Draw a decision tree to represent Westward's problem.

(b) Assuming that Westward's objective is to maximize expected profit, determine the policy that they should choose. (For simplicity, you should ignore Westward's preference for money over time: for example, the fact that they would prefer to receive a given cash inflow now rather than in the future.)

(c) In reality, Westward have little knowledge of the progress which has been made by the rival. This means that the probabilities given above for beating the rival (if the launch is, or is not, brought forward) are very rough estimates. How sensitive is the policy you identified in (b) to changes in these probabilities?

(4) The risk of flooding in land adjacent to the River Nudd has recently increased. This is because of a combination of high spring tides and the development by farmers of more efficient drainage systems in the nearby hills, which means that, after heavy rainfall, water enters the river more quickly. A tidal barrier is being constructed at the mouth of the river, but the Hartland River Authority has to decide how to provide flood protection in the two years before the barrier is completed. Flooding is only likely to occur during the spring high-tide period and the height of the river at this time cannot be predicted with any certainty. In the event of flooding occurring in any one year the Authority will have to pay out compensation of about $2 million. Currently, the Authority is considering three options.

First, it could do nothing and hope that flooding will not occur in either of the next two years. The river's natural banks will stop flooding as long as the height of the water is less than 9.5 feet. It is estimated that there is a probability of 0.37 that the height of the river will exceed this figure in any one year.

Alternatively, the Authority could erect a cheap temporary barrier to a height of 11 feet. This barrier would cost $0.9 million to erect and it is thought that there is a probability of only 0.09 that the height of the river would exceed this barrier. However, if the water did rise above the barrier in the first year, it is thought that there is a 30% chance that the barrier would be damaged, rendering it totally ineffective for the second year. The Authority would then have to decide whether to effect repairs to the barrier at a cost of $0.7 million or whether to leave the river unprotected for the second year.

The third option would involve erecting a more expensive barrier. The fixed cost of erecting this type of barrier would be $0.4 million and there would be an additional cost of $0.1 million for each foot in the barrier's height. For technical reasons, the height of this barrier would be either 11 or 13 feet, and it is thought that there would be no chance of the barrier being damaged if flooding did occur. The probability of the river's height exceeding the 13-foot barrier in any one year is estimated to be only 0.004.

(a) Draw a decision tree to represent the River Authority's problem.
(b) Determine the optimum policy for the Authority, assuming that their objective is to minimize expected costs. (For simplicity, you should ignore time preferences for money.)

(5) An enginering company is about to undertake a major overhaul of a factory's machinery for a customer. The overhaul will be carried out on a Saturday and Sunday, but, if it is not completed by the Monday morning, the factory will experience serious production losses. In this event, the engineering company has agreed to compensate the customer by paying a penalty of $20 000.

The manager of the engineering company has to decide how many engineers to include in the overhaul team. Each engineer in the team will be paid $480 for working over the weekend, but because of the nature of the work, only teams of 10, 15 or 20 engineers can be considered. The manager estimates that the chances of a 10-person team completing the overhaul by the Monday morning is only 0.4. A 15-person team has, he estimates, a 0.6 probability of meeting the deadline, while a 20-person team has a 0.9 probability of completing the work in time for Monday morning.

(a) Assuming that the manager wants to minimize expected costs, how large a team should he choose?
(b) Having made a provisional decision about the size of the team, the manager hears that a piece of specialized equipment will be available for hire on the Sunday, at short notice. The cost of hiring this equipment would be $4400, and it would require at least 15 engineers to operate it. However, it is virtually certain that the overhaul would be completed on time if the equipment was used.

Before making a decision on whether to hire the equipment, the manager will review the progress that has been made on the Saturday evening. He reckons that there is a 0.5 probability that a 15-person team would be behind schedule by Saturday evening, while there is only a 0.2 probability that a 20-person team will be in this position. He then calculates the probabilities of the overhaul overrunning the deadline if the equipment is not hired, given the position on Saturday evening. These probabilities are shown below:

	15-person team Position on Saturday evening	
	Behind Schedule	Not behind Schedule
p(overhaul exceeds deadline if equipment not hired)	0.6	0.2

	20-person team Position on Saturday evening	
	Behind Schedule	Not behind Schedule
p(overhaul exceeds deadline if equipment not hired)	0.2	0.075

How many people should the manager now include in the team and should he hire the equipment on the Saturday evening?

REFERENCES

1. Howard, R. A. (1988) Decision Analysis: Practice and Promise, *Management Science*, **34**, No. 6, 679–695.
2. Jones, J. M. (1985) Decision Analysis Using Spreadsheets, *European Journal of Operational Research*, **26**, 385–400.
3. Keefer, D. L. and Bodily, S. E. (1983) Three Point Approximations for continuous Random Variables, *Management Science*, **29**, No. 5, 595–609.
4. Pearson, E. S. and Tukey, J. W. (1965) Approximating Means and Standard Deviations Based on Distances between Percentage Points of Frequency Curves, *Biometrika*, **52**, No. 3–4, 533–546.
5. Ulvila, J. W. (1987) Postal Automation (Zip + 4) Technology: a Decision Analysis, *Interfaces*, **7**:2 1–12.
6. Cohan, D., Hass, S. M., Radloff, D. L. and Yancik, R. F. (1984) Using Fire in Forest Management: Decision Making under Uncertainty, *Interfaces*, **14**:5, 8–19.
7. Bell, D. E. (1984) Bidding for the S. S. *Kuniang*, *Interfaces*, **14**:2, 17–23.
8. Hosseini, J. (1986) Decision Analysis and its Application in the Choice between Two Wildcat Oil Ventures, *Interfaces*, **16**:2, 75–85.
9. Madden T. J., Hyrnick, M. S. and Hodde, J. A. (1983) Decision Analysis Used to Evaluate Air Quality Control Equipment for the Ohio Edison Oil Company, *Interfaces*, **13**:1, 66–75.
10. Winter, F. W. (1985) An Application of Computerized Decision Tree Models in Management–union Bargaining, *Interfaces*, **15**:2, 74–80.

11. Hertz, D. B. and Thomas, H. (1984) *Practical Risk Analysis: An Appraisal Through Case Histories*, Wiley, Chichester.
12. Von Winterfeldt, D. V. (1980) Structuring Decision Problems for Decision Analysis, *Acta Psychologica*, **45**, 73–93.
13. Fischhoff, B. (1980) Decision Analysis: Clinical Art or Clinical Science? in L. Sjoberg, T. Tyszka and J. A. Wise (eds), *Human Decision Making*, Bodafors, Doxa.
14. Keeney, R. (1980) Decision Analysis in the Geo-technical and Environmental Fields, in L. Sjoberg, T. Tyszka and J. A. Wise (eds), *Human Decision Making*, Bodafors, Doxa.
15. Humphreys, P. (1980) Decision Aids: Aiding Decisions, in L. Sjoberg, T. Tyszka and J. A. Wise (eds), *Human Decision Making*, Bodafors, Doxa.
16. Fischhoff, B., Slovic, P. and Lichtenstein, S. (1978) Fault Trees: Sensitivity of Estimated Failure Probabilities to Problem Representation, *Journal of Experimental Psychology: Human Perception and Performance*, **4**, 330–344.
17. Howard, R. A. (1989) Knowledge Maps, *Management Science*, **35**, 903–923. See also R. M. Oliver and J. Q. Smith (eds) (1990) *Influence Diagrams, Belief Nets and Decision Nets*, Wiley, Chichester.

Chapter 6

Applying Simulation to Decision Problems

INTRODUCTION

When the payoff of a decision depends upon a large number of factors, estimating a probability distribution for the possible values of this payoff can be a difficult task. Consider, for example, the problem of estimating a probability distribution for the return that might be generated by a new product. The return on the investment will depend upon factors such as the size of the market, the market share which the product will achieve, the costs of launching the product, manufacturing and distribution costs, and the life of the product. We could, of course, ask the decision maker to estimate the probability distribution directly (for example, we might ask questions such as: 'What is the probability that the investment will achieve a return of over 10% per annum?'). However, it is likely that many decision makers would have difficulty in making this sort of judgment since all the factors which might influence the return on the investment, and the large number of ways in which they could interrelate, would have to be considered at the same time.

The decision analysis approach to this problem is to help the decision maker by initially dividing the probability assessment task into smaller parts (a process sometimes referred to as 'credence decomposition'). Thus we might ask the decision maker to estimate individual probability distributions for the size of the market, the market share which will be achieved, the launch costs and so on.

The problem is, that having elicited these distributions, we then need to determine their combined effect in order to obtain a probability distribution for the return on the investment. In most practical problems there will be a large number of factors, and also the possible values which each of the factors can assume may be very large or infinite. Consider, for example, the possible levels of the costs of launch, manufacturing and distribution which we might experience. All of this means that there will be a large or infinite number of combinations of circumstances which could affect the return on the investment. In such situations it is clearly impractical to use an approach such as a probability tree to calculate the probability of each of these combinations of circumstances occurring.

One answer to our problem is to use a versatile and easily understood technique called Monte Carlo simulation. This involves the use of a computer to generate a large number of possible combinations of circumstances which might occur if a particular course of action is chosen. When the simulation is performed the more likely combination of circumstances will be generated most often while very unlikely combinations will rarely be generated. For each combination the payoff which the decision maker would receive is calculated and, by counting the frequency with which a particular payoff occurred in the simulation, a decision maker is able to estimate the probability that the payoff will be received. Because this method also enables the risk associated with a course of action to be assessed it is often referred to as *risk analysis* (although some authors use the term to include methods other than simulation such as mathematical assessment of risk). Monte Carlo simulation is demonstrated in the next section.

MONTE CARLO SIMULATION

As we stated earlier, a computer is normally used to carry out Monte Carlo simulation, but we will use the following simplified problem to illustrate how the technique works. A company accountant has estimated that the following probability distributions apply to his company's inflows and outflows of cash for the coming month:

Cash inflows ($)	Probability (%)	Cash outflows ($)	Probability (%)
50 000	30	50 000	45
60 000	40	70 000	55
70 000	30		——
	——		100
	100		

The accountant would like to obtain a probability distribution for the net cash flow (i.e. cash inflow − cash outflow) over the month. He thinks that it is reasonable to assume that the outflows and inflows are independent.

Of course, for a simple problem like this we could obtain the required probability distribution by calculation. For example, we could use a probability tree to represent the six combinations of inflows and outflows and then calculate the probability of each combination occurring. However, since most practical problems are more complex than this we will use the example to illustrate the simulation approach. The fact that we can calculate the probabilities exactly for this problem will have the advantage of enabling us to assess the reliability of estimates which are derived from simulation.

In order to carry out our simulation of the company's cash flows we will make use of *random numbers*. These are numbers which are produced in a manner analogous to those that would be generated by spinning a roulette wheel (hence the name Monte Carlo simulation). Each number in a specified range (e.g. 00–99) is given an equal chance of being generated at any one time. In practical simulations, random numbers are normally produced by a computer, although, strictly speaking, most computers generate what are referred to as pseudo-random numbers, because the numbers only have the appearance of being random. If you had access to the computer program which is being used to produce the numbers and the initial value (or seed) then you would be able to predict exactly the series of numbers which was about to be generated.

Before we can start our simulation we need to assign random numbers to the different cash flows so that, once a particular random number has been generated, we can determine the cash flow which it implies. In this example we will be using the hundred random numbers between 00 and 99. For the cash inflow distribution we therefore assign the random numbers 00 to 29 (30 numbers in all) to an inflow of $50 000 which has a 30% probability of occurring. Thus the probability of a number between 00 and 29 being generated mirrors exactly the probability of the $50 000 cash inflow occurring. Similarly, we assign the next 40 random numbers: 30 to 69 to a cash inflow of $60 000, and so on, until all 100 numbers have been allocated.

Cash inflow ($)	Probability (%)	Random numbers
50 000	30	00–29
60 000	40	30–69
70 000	30	70–99

The process is repeated for the cash outflow distribution and the allocations are shown below.

Cash outflow ($)	Probability (%)	Random numbers
50 000	45	00–44
70 000	55	45–99

We are now ready to perform the simulation run. Each simulation will involve the generation of two random numbers. The first of these will be used to determine the cash inflow and the second, the cash outflow. Suppose that a computer generates the random numbers 46 and 81. This implies a cash inflow of $60 000 and an outflow of $70 000, leading to a net cash flow for the month of −$10 000. If we repeat this process a large number of times then it is to be expected that the more likely combinations of cash flows will occur most often, while the unlikely combinations will occur relatively infrequently. Thus the probability of a particular net cash flow occurring can be estimated from the frequency with which it occurs in the simulations. Table 6.1 shows the results of ten simulations. This number of repetitions is far too small for practical purposes, but the experiment is designed simply to illustrate the basic idea.

Table 6.1 Ten simulations of monthly cash flows

Random number	Cash inflow ($)	Random number	Cash outflow ($)	Net cash flow ($)
46	60 000	81	70 000	− 10 000
30	60 000	08	50 000	10 000
14	50 000	88	70 000	− 20 000
35	60 000	21	50 000	10 000
09	50 000	73	70 000	− 20 000
19	50 000	77	70 000	− 20 000
72	70 000	01	50 000	20 000
20	50 000	46	70 000	− 20 000
75	70 000	97	70 000	0
16	50 000	43	50 000	0

If we assume for the moment that this small number of repetitions is sufficient to give us estimates of the probabilities of the various net cash flows occurring then we can derive the probability distribution shown in Table 6.2. For example, since a net cash flow of −$20 000 occurred in four of our ten simulations we estimate that the probability of this net cash flow occurring is 4/10. Note that the table also shows the probability distribution which we would have derived if we had used a probability tree to calculate the probabilities. The discrepancies between the two distributions show that the result based on only ten simulations gives a poor estimate of the real distribution. However, as more simulations are carried out we can expect this estimate to improve. This is shown in Table 6.3, which compares the 'real' distribution with estimates based on 50, 1000 and 5000 simulations which were carried out on a microcomputer.

How many simulations are needed to give an acceptable level of reliability? This question can be answered by using relatively complex iterative statistical

Table 6.2 Estimating probabilities from the simulation results

Net cash flow ($)	Number of simulations resulting in this net cash flow	Probability estimate based on simulation	Calculated probability
−20 000	4	4/10 = 0.4	0.165
−10 000	1	1/10 = 0.1	0.220
0	2	2/10 = 0.2	0.300
10 000	2	2/10 = 0.2	0.180
20 000	1	1/10 = 0.1	0.135

Table 6.3 The effect of the number of simulations on the reliability of the probability estimates

Net cash flow ($)	Probability estimates based on:		5000 simulations	Calculated probability
	50	1000		
−20 000	0.14	0.164	0.165	0.165
−10 000	0.18	0.227	0.216	0.220
0	0.42	0.303	0.299	0.300
10 000	0.12	0.168	0.184	0.180
20 000	0.14	0.138	0.136	0.135

methods, but a simpler approach is to start off with a run of about 250 simulations and then increase the length of the runs until there is virtually no change in the estimates produced by the simulation. It is unlikely that runs of more than 1000 simulations will be required.

APPLYING SIMULATION TO A DECISION PROBLEM

The Elite Pottery Company

We will now show how simulation can be applied to a decision problem. The Elite Pottery Company is planning to market a special product to commemorate a major sporting event which is due to take place in a few months' time. A large number of possible products have been considered, but the list has now been winnowed down to two alternatives: a commemorative plate and a figurine. In order to make a decision between the plate and the figurine the company's managing director needs to estimate the profit which would be earned by each product (the decision will be made solely on the basis of profit so that other objectives such as achieving a prestigious company image, increasing public awareness of the company, etc. are not considered to be important). There is some uncertainty about the costs of manufacturing the products and the levels of sales, although it is thought that all sales will be made in the very short period which coincides with the sporting event.

The application of simulation to a problem like this involves the following stages:

1. Identify the factors that will affect the payoffs of each course of action.
2. Formulate a model to show how the factors are related.
3. Carry out a preliminary sensitivity analysis to establish the factors for which probability distributions should be assessed.
4. Assess probability distributions for the factors which were identified in stage 3.
5. Perform the simulation.
6. Apply sensitivity analysis to the results of the simulation.
7. Compare the simulation results for the alternative courses of action and use these to identify the preferred course of action.

We will now show how each of these stages can be applied to the Elite Pottery Company problem.

Stage 1: Identify the factors

The first stage in our analysis of Elite's problem is to identify the factors which it is thought will affect the profit of the two products. For brevity, we will only consider in detail the factors affecting the potential profit of the commemorative plate. A tree diagram may be helpful in identifying these factors, since it enables them to be sub-divided until the decision maker feels able to give a probability distribution for the possible values which the factor might assume. Figure 6.1 shows a tree for this problem. Of course, it might have been necessary to extend the tree further, perhaps by sub-dividing fixed costs into different types, such as advertising and production set-up costs or breaking sales into home and export sales. It is also worth noting that subsequent analysis will be simplified if the factors can be identified in such a way that their probability distributions can be considered to be independent. For example, we will assume here that variable costs and sales are independent. However, in practice, there might be a high probability of experiencing a variable cost of around $7 per plate if less than 10 000 plates are sold, while we might expect costs of around $5 if more than 10 000 are sold because of savings resulting from the bulk purchase of materials, etc. It is possible to handle dependence, as we will show later, but it does add complications to the analysis.

Stage 2: Formulate a model

Having identified the factors for which probability distributions can be assessed, the next step is to formulate a mathematical model to show how

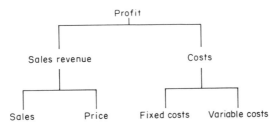

Figure 6.1 Identifying the factors which will affect the profit earned by the commemorative plate

they affect the variable of interest, in this case profit. For Elite's problem the following model is thought to be appropriate:

$$\text{Profit} = (\text{Price} - \text{Variable cost}) \times \text{Sales} - \text{Fixed costs}$$

Of course, this is only a model and therefore almost certainly a simplification of the real problem. In practice, a large number of factors and relationships which we have not included in the model may affect profit. For example, at higher sales levels more breakages may occur because of increased pressure on the workforce. Similarly, fixed costs may only remain at a particular level up to a certain number of sales when new equipment and machinery may be required. However, if we tried to include every possible factor and relationship in the model it would become too complicated to handle and the valuable insights which can be gained by building and analyzing the model might be lost. Therefore a balance has to be struck between the need to keep the model simple and understandable and the need to provide a reasonable and plausible representation of the real problem.

Stage 3: Preliminary Sensitivity Analysis

Although we may be uncertain about the values which the factors might assume, this uncertainty may not be important in the case of all the factors. For example, we might find that there would be little change in profit if fixed costs changed from their lowest to their highest possible value. If this were the case then it would clearly be a waste of effort if time was spent in eliciting a probability distribution for fixed costs: the use of a single figure representing the most likely value would suffice. Sensitivity analysis can therefore be helpful at this stage in screening out those factors which do not require a probability distribution. This analysis can be carried out as follows:

(i) Identify the lowest, highest and most likely values that each factor can assume.

The values put forward for the factors in Elite's problem are shown in Table 6.4. (It is assumed that the price of the plate will be fixed at $25 so there is no uncertainty associated with this factor.)

Table 6.4 Estimates of lowest, highest and most likely values for the Elite Pottery problem

Factor	Most likely value	Lowest possible value	Highest possible value
Variable costs	$13	$8	$18
Sales	22 000 units	10 000 units	30 000 units
Fixed costs	$175 000	$100 000	$300 000

(ii) Calculate the profit which would be achieved if the first factor was at its lowest value and the remaining factors were at their most likely values.

Thus, if variable costs are at their lowest possible value of $8 and the remaining factors are at their most likely values we have:

$$\text{Profit} = (\$25 - \$8) \times 22\,000 - \$175\,000 = \$199\,000$$

(iii) Repeat (ii), but with the first factor at its highest possible value. Thus we have:

$$\text{Profit} = (\$25 - \$18) \times 22\,000 - \$175\,000 = -\$21\,000$$

(iv) Repeat stages (ii) and (iii) by varying, in turn, each of the other factors between their lowest and highest possible values while the remaining factors remain at their most likely values.

Figure 6.2 shows the results of the preliminary sensitivity analysis. This figure indicates that each of the factors is crucial to our analysis in that a change from its lowest to its highest possible value will have a major effect on profit. It is therefore well worth spending time in the careful elicitation of probability distributions for all these factors.

Stage 4: Assess Probability Distributions

Figure 6.3 shows the probability distributions which were obtained for variable costs, sales and fixed costs. Techniques for eliciting probability distributions are discussed in Chapter 10.

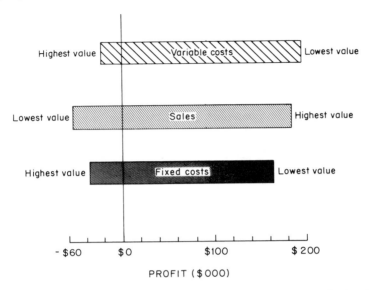

Figure 6.2 Preliminary sensitivity analysis showing the effect on profit if each factor changes from its lowest to its highest possible value

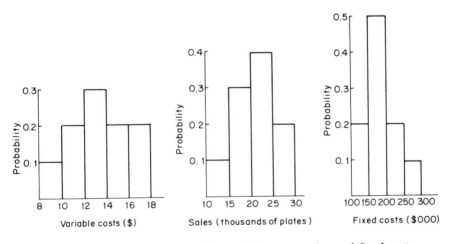

Figure 6.3 Probability distributions for variable costs, sales and fixed costs

Stage 5: Perform the simulation

Simulation can now be used to obtain a probability distribution for the profit which the plate will earn. A microcomputer was programmed to carry out the simulation which involved the generation of three random numbers. The first was used to give a figure for variable costs, the second a figure for sales and the third a figure for fixed costs. For example, in the first simulation the computer produced variable costs of $13.2, sales of 26 500 and fixed costs of $125 000, and the resulting profit was calculated as follows:

$$\text{Profit} = (\$25 - \$13.2) \times 26\,500 - \$125\,000 = \$187\,700$$

This process was then repeated until 500 simulations had been carried out. The profits generated by these simulations and the resulting probability distribution are shown below:

Profit ($)		No. of simulations	Probability
−200 000 to under	−100 000	26	26/500 = 0.052
−100 000 to under	0	120	120/500 = 0.240
0 to under	100 000	213	213/500 = 0.426
100 000 to under	200 000	104	104/500 = 0.208
200 000 to under	300 000	34	34/500 = 0.068
300 000 to under	400 000	3	3/500 = 0.006
		500	1.000

Mean profit = $51 800

This distribution is illustrated in Figure 6.4. It can be seen that it is likely that profit will fall between $0 and $100 000. There is, however, a probability of 0.292 that the product will make a loss and it is unlikely that profits will exceed $200 000.

Stage 6: Sensitivity Analysis on the Results of the Simulation

Hertz and Thomas[1] argue that Monte Carlo simulation is itself a comprehensive form of sensitivity analysis so that, in general, further sensitivity tests are not required. However, if the decision maker has some doubts about the probability distributions which were elicited from him then the effects of changes in these distributions on the simulation results can be examined. There are several ways in which this sensitivity analysis can be carried out (see for example, Singh and Kiangi[2]). It might simply involve changing the distributions, repeating the simulation and examining the resulting changes on the profit probability distribution. Similarly, if the decision maker is not confident that the structure of the model is correct then the effect of changes in this can be examined. Clearly, if such changes have minor effects on the simulation results then the original model can be assumed to be adequate.

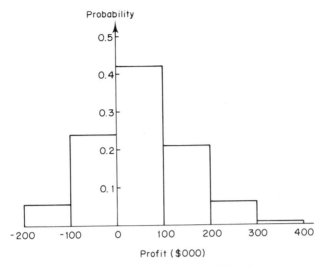

Profit ($000)

Figure 6.4 Probability distribution for profit earned by the commemorative plate

Stage 7: Compare Alternative Course of Action

You will recall that Elite Pottery had to make a decision between the production of the commemorative plate and the figurine. The factors which it was thought would affect the profit on the figurine were also identified and a simulation carried out. This resulted in the probability distribution shown below:

Profit on figurine ($)	Probability
−300 000 to under −200 000	0.06
−200 000 to under −100 000	0.10
−100 000 to under 0	0.15
0 to under 100 000	0.34
100 000 to under 200 000	0.18
200 000 to under 300 000	0.08
300 000 to under 400 000	0.05
400 000 to under 500 000	0.04
	1.00

Mean profit = $62 000

In order to help him to choose between the products the decision maker clearly needs to compare the two profit probability distributions. This comparison can be made in a number of ways.

Plotting the Two Distributions

By inspecting graphs of the two probability distributions the decision maker can compare the probabilities of each product making a loss or the probabilities that each product would reach a target level of profit. The two distributions have been plotted in Figure 6.5. Note that to make the comparison between the distributions easier, their histograms have been approximated by line graphs (or polygons). Although both distributions have their highest probabilities for profits in the range $0 to under $100 000 it can be seen that the distribution for the figurine has a greater spread. Thus while the figurine has a higher probability of yielding a large loss it also has a higher probability of generating a large profit. Clearly, the greater spread of the figurine's distribution implies that there is more uncertainty about the profit which will actually be achieved. Because of this, the spread of a distribution is often used as a measure of the risk which is associated with a course of action.

A distribution's spread can be measured by calculating its *standard deviation* (the larger the standard deviation, the greater the spread—see the appendix to this chapter if you are unfamiliar with this measure). For the plate, the standard deviation of profits is $99 080 while for the figurine it is $163 270.

Determining the Option with the Highest Expected Utility

Because the two products offer different levels of risk, utility theory can be used to identify the option which the decision maker should choose. The

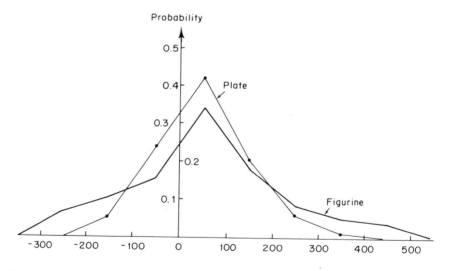

Figure 6.5 A comparison of the profit probability distributions of the commemorative plate and the figurine

problem is, unlike the examples which we encountered in Chapter 4, each option has a very large number of possible outcomes. One way around this problem is to find a mathematical function which will approximate the decision maker's utility function. A computer can then be used to calculate the utility for each profit generated in the simulation run. The resulting utilities would then be averaged to give the expected utility.

Stochastic Dominance

Sometimes the alternative with the highest expected utility can be identified by a short-cut method which is based on a concept known as stochastic dominance. This exists where the expected utility of one option is greater than that of another for an entire class of utility functions. This means that we can be sure that the option will be preferred without going to the trouble of eliciting the decision maker's complete utility function; all we need to establish is that the utility function has some basic characteristics.[3]

Stochastic dominance can be recognized by plotting the cumulative probability distribution functions (CDFs). As we saw in Chapter 3, the CDF shows the probability that a variable will have a value less than any given value. First- and second-degree stochastic dominance are the two most useful forms which the CDFs can reveal.

First-degree Stochastic Dominance

This concept requires some very unrestrictive assumptions about the nature of the decision maker's utility function. When money is the attribute under consideration, the main assumption is simply that higher monetary values have a higher utility. To illustrate the application of first-degree stochastic dominance, consider the following simulation results which relate to the profits of two potential products, P and Q:

Product P			Product Q		
Profit ($m)	Prob.	Cumulative prob.	Profit ($m)	Prob.	Cumulative prob.
0 to under 5	0.2	0.2	0 to under 5	0	0
5 to under 10	0.3	0.5	5 to under 10	0.1	0.1
10 to under 15	0.4	0.9	10 to under 15	0.5	0.6
15 to under 20	0.1	1.0	15 to under 20	0.3	0.9
			20 to under 25	0.1	1.0

The CDFs for the two products are plotted in Figure 6.6. It can be seen the CDF for product Q is always to the right of that for product P. This means that for any level of profit, Q offers the smallest probability of falling below

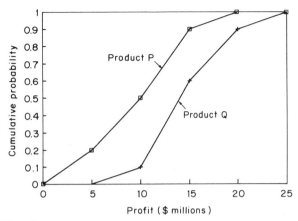

Figure 6.6 First-degree stochastic dominance

that profit. For example, Q has only a 0.1 probability of yielding a profit of less than $10 million while there is a 0.5 probability that P's profit will fall below this level. Because Q's CDF is always to the right of P's, we can say that Q exhibits first-degree stochastic dominance over P. Thus, as long as the weak assumptions required by first-degree stochastic dominance apply, we can infer that product Q has the highest expected utility.

Second-degree Stochastic Dominance

When the CDFs for the options intersect each other at least once it may still be possible to identify the preferred option if, in addition to the weak assumptions we made for first-degree stochastic dominance, we can also assume that the decision maker is risk averse (i.e. his utility function is concave) for the range of values under consideration. If this assumption is appropriate then we can make use of second-degree stochastic dominance. To demonstrate the application of this concept let us compare the following simulation results which have been generated for two other potential products, R and S:

Product R			*Product S*		
Profit ($m)	*Prob.*	*Cumulative prob.*	*Profit ($m)*	*Prob.*	*Cumulative prob.*
0 to under 5	0.1	0.1	0 to under 5	0.3	0.3
5 to under 10	0.3	0.4	5 to under 10	0.3	0.6
10 to under 15	0.4	0.8	10 to under 15	0.2	0.8
15 to under 20	0.2	1.0	15 to under 20	0.1	0.9
20 to under 25	0	1.0	20 to under 25	0.1	1.0

The CDFs for the two products are shown in Figure 6.7. It can be seen that for profits between $0 and $15 million, R is the dominant product, while for the range $15–25 million, S dominates. To determine which is the dominant product overall we need to compare both the lengths of the ranges for which the products are dominant and the extent to which they are dominant within these ranges (i.e. the extent to which one curve falls below the other). This comparison can be made by comparing area X, which shows the extent to which R dominates S, with area Y, the extent to which S dominates R. Since, area X is larger, we can say that product R has second-degree stochastic dominance over product S. Again, as long as our limited assumptions about the form of the decision maker's utility function are correct then we can conclude that R has a higher expected utility than S.

Of course, there are bound to be situations where the CDFs intersect each other several times. In these cases we would have to add areas together to establish the extent to which one option dominates the other.

The Mean–Standard Deviation Approach

When a decision problem involves a large number of alternative courses of action it is helpful if inferior options can be screened out at an early stage. In these situations the mean–standard deviation approach can be useful. This has mainly been developed in connection with portfolio theory, where a risk-averse decision maker has to choose between a large number of possible investment portfolios (see Markowitz[4]).

To illustrate the approach let us suppose that a company is considering five alternative products which are code-named A to E. For each product

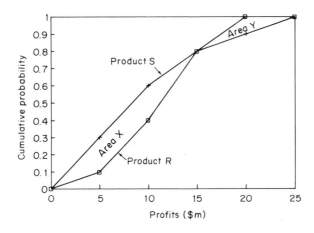

Figure 6.7 Second-degree stochastic dominance

a simulation has been carried out and the mean and standard deviation of that product's profits have been calculated. The results are plotted in Figure 6.8. The company's manager would like to maximize the expected (or mean) return, and being risk averse, she would also like to minimize the risk or uncertainty which her company faces. If we compare products A and B we see that, while they offer the same expected return, product B is much more risky. Product A is therefore said to dominate B. B is also dominated by C, which for the same level of risk offers higher expected profits. For the same reason, D dominates E. The non-dominated products, A, C and D, are therefore said to lie on the efficient frontier, and only these products would survive the screening process and be considered further. The choice between A, C and D will depend on how risk averse the decision maker is. Product A offers a low expected return but also a low level of risk, while at the other extreme, C offers high expected returns but a high level of risk. The utility approach, which we discussed above, could now be used to compare these three products.

Note that for the mean–standard deviation screening process to be valid it can be shown that a number of assumptions need to be made. First, the probability distributions for profit should be fairly close to the normal distribution shape shown in Figure 6.9(a) (in many practical situations this is likely to be the case: see Hertz and Thomas[1]). Second, the decision maker

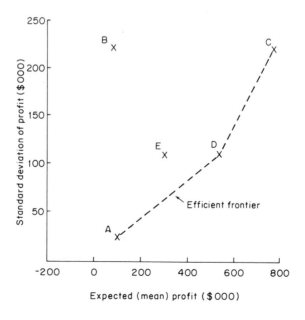

Figure 6.8 The mean-standard deviation screening method

should have a utility function which not only indicates risk aversion but which also has (at least approximately) a quadratic form. This means that the function can be represented by an equation of the form:

$$U(x) = c + bx + ax^2$$

where x = a given sum of money, $U(x)$ = the utility of this sum of money and a, b and c are other numbers which determine the exact nature of the utility function.

For example, Figure 6.9(b) shows the utility functions

$$U(x) = 0 + 0.4x - 0.04x^2$$

and

$$U(x) = 0 + 0.25x - 0.01x^2$$

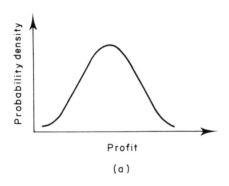

Profit

(a)

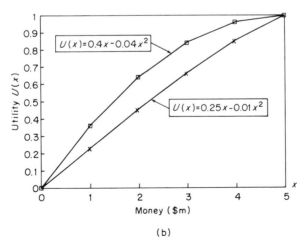

(b)

Figure 6.9 (a) A normal probability distribution for profit; (b) examples of quadratic utility functions

(where x = monetary sums in millions of dollars) for monetary values between $0 and $5 million. Of course, not all utility functions have the quadratic form, but as Markowitz[4] argues, 'the quadratic nevertheless shows a surprising flexibility in approximating smooth, concave curves'.

Having considered the different ways in which the results of simulations can be compared, how was a decision made in the case of the Elite Pottery Company? First, it was established that the Managing Director's utility function for profit was concave (i.e. he was risk averse). Second, a plot of the cumulative probability distributions for the profits earned by the plate and the figurine showed that the figurine had second-degree stochastic dominance over the plate. Thus while the figurine had a slightly higher probability of yielding a loss, it could also generate much larger profits than the plate, and this was sufficient for it to have the highest expected utility, even though the manager was risk averse. A decision was therefore made to go ahead with production of the figurine.

APPLYING SIMULATION TO INVESTMENT DECISIONS

The techniques which we have outlined in this chapter have been most widely applied in the area of investment appraisal. In this section we will briefly discuss the commonly used net present value (NPV) approach to investment appraisal and then show how Monte Carlo simulation can be used to improve on the 'standard' NPV method, which is based on single-figure estimates of cash flows. We also show that the NPV method makes some very strong assumptions about the nature of the decision maker's preferences, which in some circumstances may not be justified.

The Net Present Value (NPV) Method

Our intention here is to give only a brief overview of the net present value method. More detailed explanations can be found in accountancy textbooks (e.g. Weston and Brigham[5] and Drury[6]).

When money is invested in a project a commitment of funds is generally required immediately. However, the flow of funds earned by the investment will occur at various points of time in the future. Clearly, receiving $1000 in, say, a year's time is less attractive than receiving $1000 now. The $1000 received now could be invested, so that in a year's time it will have earned interest. Similarly, $1000 due to be received in 2 years' time will be less attractive than $1000 which will be received one year from now. This implies that money which will be earned in the future should be *discounted* so that its value can be compared with sums of money which are being held now. The process involved is referred to as 'discounting to present value'. For example, we might judge that the $1000 due in one year is only equivalent to receiving $909 now, while the $1000 due in 2 years has only the same value as receiving $826 now.

The severity with which we discount future sums of money to their present value is reflected in the discount rate. Determining the appropriate discount rate for a company's potential investment projects is, ultimately, a matter of judgment and preference. However, many attempts have been made to make the choice of a discount rate as 'objective' as possible, making this a complex area which is beyond the scope of this text. For many situations, it will be convenient to let the discount rate reflect the opportunity cost of the capital which is being invested (i.e. the rate of return which could be earned on the best alternative investment). Thus if we are only considering two mutually exclusive projects A and B and we could earn a 12% return on project A, then the discount rate for project B would be 12% because, if we invest in B, we will be forgoing the 12% return which A would have generated. Having determined the appropriate discount rate, the process of discounting future sums of money is very straightforward. It simply involves multiplying the sum of money by a *present value factor*, which can be obtained from published tables.

Let us now use a simple example to illustrate the net present value approach to investment appraisal. A company has to choose between two new machines, the Alpha and the Beta. Both machines would cost $30 000 and have an expected lifetime of 4 years. Estimates of the annual cash inflows which each machine would generate are given below together with estimates of the cash outflows which would be experienced for each year of the machine's operation. For simplicity, we will assume that all the cash flows occur at the year end.

	Alpha machine			
Time of cash flow	Year 1	Year 2	Year 3	Year 4
Cash inflows	$14 000	$15 000	$15 000	$14 000
Cash outflows	$2 000	$4 000	$6 000	$7 000

	Beta machine			
Time of cash flow	Year 1	Year 2	Year 3	Year 4
Cash inflows	$8 000	$13 000	$15 000	$21 500
Cash outflows	$4 000	$4 000	$5 000	$5 000

Table 6.5 shows the calculations which are involved in determining the net present value of the two potential investments. First, the net cash flow is determined for each year. These net cash flows are then discounted by multiplying by the appropriate present value factor. (The present value factors used in Table 6.5 are based on the assumption that a 10% discount rate

is appropriate.) Finally, these discounted cash flows are summed to give the net present value of the project. It can be seen that, according to the NPV criterion, the Alpha machine offers the most attractive investment opportunity.

Table 6.5 Calculating the NPVs for the Alpha and Beta machines

Time of cash flow	Cash inflow ($)	Cash outflow ($)	Net cash flow ($)	Present value factor	Discounted cash flow ($)
(a) *Alpha machine*					
Now	0	30 000	−30 000	1.0000	−30 000
Year 1	14 000	2 000	12 000	0.9091	10 909
Year 2	15 000	4 000	11 000	0.8264	9 090
Year 3	15 000	6 000	9 000	0.7513	6 762
Year 4	14 000	7 000	7 000	0.6830	4 781
			Net present value (NPV) =		$1 542
(b) *Beta machine*					
Now	0	30 000	−30 000	1.0000	−30 000
Year 1	8 000	4 000	4 000	0.9091	3 636
Year 2	13 000	4 000	9 000	0.8264	7 438
Year 3	15 000	5 000	10 000	0.7513	7 513
Year 4	21 500	5 000	16 500	0.6830	11 270
			Net present value (NPV) =		−$143

While this approach to investment appraisal is widely used, the NPV figures are obviously only as good as the estimates on which the calculations are based. In general, there will be uncertainty about the size of the future cash flows and about the lifetime of the project. Expressing the cash flow estimates as single figures creates an illusion of accuracy, but it also means that we have no idea as to how reliable the resulting NPV is. For example, it may be that the year 1 cash inflow for the Beta machine could be anything from $2000 to $14 000, and we have simply used the mid-range figure, $8000, as our estimate. If the actual flow did turn out to be near $14 000 then our estimated NPV would have been very misleading.

Clearly, the approach would be more realistic if we could incorporate our uncertainty about the cash flows into the analysis. The result would be a probability distribution for the NPV which would indicate the range within which it would be likely to lie and the probability of it having particular values. From this we could assess the chances of the project producing a negative NPV or the probability that the NPV from one project will exceed that of a competing project. In the section which follows we will show how simulation can be used to extend the NPV approach so that we can explicitly take into account the uncertainty associated with an investment project.

Using Simulation

Let us first consider the application of simulation to the Alpha machine. It was thought that the following factors would affect the return on this investment:

(i) The price of the machine;
(ii) The revenue resulting from the machine's output in years 1 to 4;
(iii) Maintenance costs in years 1 to 4;
(iv) The scrap value of the machine at the end of year 4.

The price of the machine was known to be $30 000, but because there was uncertainty about the other factors, probability distributions were elicited from the company's management. The shapes of these distributions are shown in Figure 6.10 (for simplicity, it was assumed that the distributions were independent).

Random numbers were then used to sample a value from each distribution, using the methods which we outlined earlier, and the NPV was calculated for this set of values. For example, the first simulation generated the following values:

			Purchase price = $30 000
Year 1 Revenue	= $24 500	Maintenance costs =	$2 150
Year 2 Revenue	= $14 200	Maintenance costs =	$3 820
Year 3 Revenue	= $17 320	Maintenance costs =	$4 340
Year 4 Revenue	= $16 970	Maintenance costs =	$9 090
	Scrap value = $1 860		

This led to an NPV of $8 328. The process was then repeated until 500 simulations had been carried out. Figure 6.11 shows the resulting probability distribution for the net present value. It can be seen that the NPV could be between about −$20 000 and $36 000. Moreover, although the expected (mean) NPV was $7364, there was roughly a 20% probability that the NPV would be negative.

A similar simulation was carried out for the Beta machine and the resulting distribution is also displayed in Figure 6.11. While this machine has a higher probability (about 30%) of yielding a negative NPV, its distribution is much tighter than that of the Alpha machine. For example, in contrast to the Alpha machine, there is little chance of the NPV being below −$5000.

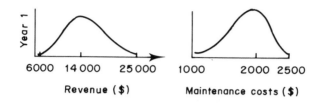

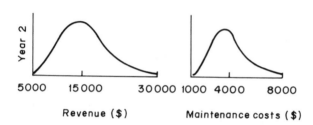

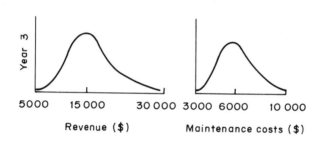

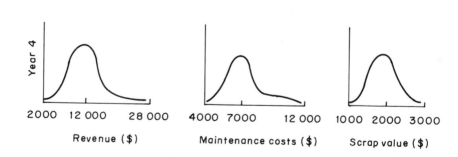

Figure 6.10 Probability distributions for the Alpha machine (vertical axes represent probability density)

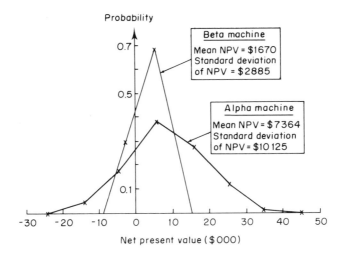

Figure 6.11 Probability distributions for the NPVs of the Alpha and Beta machines

It can be seen from these results that simulation enables a more informed choice to be made between investment opportunities. By restricting us to single-value estimates the conventional NPV approach yields no information on the level of uncertainty which is associated with different options. Hespos and Strassman[7] have shown how the simulation approach can be extended to handle investment problems involving sequences of decisions using a method known as stochastic decision tree analysis.

Utility and Net Present Value

In order to help a decision maker to choose between a number of alternative investment projects we could obtain a utility function for net present value. This would involve giving the highest possible NPV a utility of 1, the lowest a utility of 0 and the use of questions involving lotteries to determine the utilities of intermediate NPVs. The decision maker would then be advised to choose the investment project which had the highest expected utility. The question is, how justified would we be in converting NPVs to utilities? As we demonstrate below, to use NPVs we need to make some strong assumptions about the decision maker's preferences.

First, the use of the NPV approach implies that the decision maker's relative strength of preference for receiving money in any two adjacent years is the same, whatever those years are. For example, if a 10% discount rate is used it implies that $1 receivable in one year's time is equivalent to receiving about $0.91 today. Similarly, $1 receivable in 10 years' time is equivalent to receiving $0.91 in 9 years' time. Now it may well be that a decision maker has a very strong relative preference for receiving money now rather than in a year's time, while his or her relative preference between receiving a sum in 9 as opposed to 10 years is much weaker.

Second, if we consider the decision maker's relative preference for sums of money between the *same* pair of years it can be seen that the NPV method assumes a constant rate of trade-off between the years. For example (again assuming a 10% discount rate), it assumes that you would be prepared to give up the promise of $1 in a year's time in order to receive $0.91 today, and that you would be prepared to go on making this exchange, irrespective of the amount of money which is transferred from next year to this. Again, this may not be the case. You may be desperate for $10 000 now and be prepared to give up the promise of $3 in a year's time for each $0.91 you can receive today. Once you have obtained your $10 000, however, your relative preference for money received now may decline, and you may then only be prepared to give up the promise of $1 for each $0.91 you can receive now.

Clearly, if either of the NPV assumptions is *seriously* violated then the NPV will not accurately represent the decision maker's preferences between sums of money arriving at different points in time. In this case, converting the NPVs to utilities might not lead to a ranking of investment options, which reflects the decision maker's true preferences. It is therefore worth questioning whether the implicit assumptions of NPV are reasonable before applying the method.

MODELING DEPENDENCE RELATIONSHIPS

So far we have assumed, for simplicity, that all the probability distributions in our models are independent. In reality, it is possible that the value of some variables will depend upon the value of others. For example, in the Alpha and Beta machine problem it is possible that maintenance costs will be related to sales revenue, since high sales revenue implies high production levels and hence more wear and tear on machinery. Similarly, the year 2 sales revenue may be closely related to that achieved in year 1 since, for example, high sales in year 1 may signify that the product is popular and hence increase the probability of high sales in subsequent years. In order to indicate how these relationships can be modeled we need to address two issues. First, we need to consider the problems a decision maker might have in judging the extent to which one variable will depend on another. Then we need to show how the simulation can be operated so that the dependency is taken into account in the model.

Judgmental Problems

There is some evidence that decision makers have difficulties in accurately assessing the strength of association between variables. For example, Chapman and Chapman[8] described a phenomenon which they referred to as illusory correlation. In their experiment naive judges were given information on several hypothetical mental patients. This information consisted of a diagnosis and a drawing made by the patient of a person. Later the judges were asked to estimate how frequently certain characteristics referred to in the diagnosis, such as suspiciousness, had been accompanied by features of the drawing, such as peculiar eyes. It was found that the judges significantly overestimated the frequency with which, for example, suspiciousness and peculiar eyes had occurred together. Moreover, this illusory correlation survived even when contradictory evidence was presented to the judges. Tversky and Kahneman[9] have suggested that such biases are a consequence of the availability heuristic (see Chapter 8). It is easy to imagine a suspicious person drawing a person with peculiar eyes, and because of this the real frequency with which the factors co-occurred was grossly overestimated. This research indicates the powerful and persistent influence which preconceived notions can have on judgments about relationships.

As we saw in Chapter 3, the dependency between one variable and another can be expressed as a conditional probability. We might, for example, ask a decision maker to estimate a probability distribution for sales in year 2 *given that* sales in year 1 reached 20 000 units. Again there is some research evidence that such estimates may be biased. Consider, for example, the problem of estimating the probability that a patient has a disease given that a particular symptom is present. Suppose that you are presented with the information below which is based on the records of 27 patients:

	Patient: (No. of patients)	
	Has illness	*Does not have illness*
Symptom present	12	6
Symptom absent	6	3

Would you conclude that there is a relationship between the symptom and the disease? Research by Arkes *et al.*[10] suggests that many people would conclude that there was. Yet if we calculate the conditional probabilities we find:

$$p(\text{patient has illness}|\text{symptom present}) = 12/18 = 2/3$$

and

$$p(\text{patient has illness}|\text{symptom absent}) = 6/9 = 2/3$$

which shows that the presence or absence of the symptom has no effect on the probability of having the illness, so there is no relationship present. This study and others (e.g. Smedslund[11]) suggest that people only consider the figure in the top left-hand corner cell of the above table, that is, they focus only the patients having the symptom and the disease. The large value in this cell in the example above creates the illusion of a relationship. Data in the other cells (eg. the number of patients having the symptom but not the illness, or the number of patients having the illness but not the symptom) were ignored, yet they are crucial to any accurate assessment of the strength of the relationship. This suggests that people tend to focus on, or recall, the occasions when events occur together, thereby forgetting instances when there was no co-occurrence. Again, this implies that there is a danger of assuming dependency where none exists, or of overestimating the degree of dependence.

Simulating Dependence Relationships

Before we discuss methods for simulating relationships between variables, we need to consider the forms these relationships might take. One way of displaying the relationship between two variables is to draw a scatter diagram. To do this, we first need to identify the independent and dependent variable in the relationship. For example, if we think that a company's sales revenue depends largely on the amount it spends on advertising, then sales

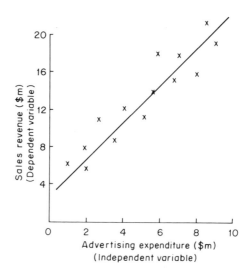

Figure 6.12 A scatter diagram showing the association between advertising expenditure and sales revenue

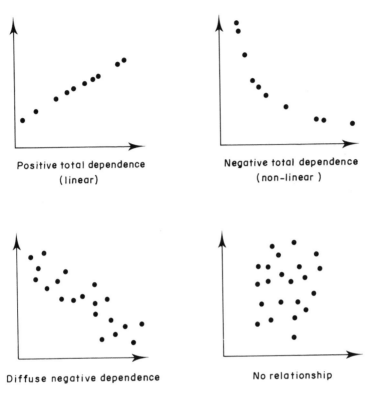

Figure 6.13   Some typical patterns which can occur on scatter diagrams

revenue will be the dependent variable and advertising expenditure the independent variable. The vertical axis of a scatter diagram is then used to represent values of the dependent variable while the horizontal axis represents the independent variable. Having drawn the axes of the diagram, we simply plot points to represent the pairs of values. Thus we might find that when we spent $1 million on advertising our sales were worth $6 million, while when $4 million was spent on advertising, sales reached $12 milion. These points and others are plotted in Figure 6.12.

The pattern revealed by this scatter diagram has three characteristics. First, it is generally the case that greater advertising is associated with higher sales. This is therefore a positive relationship. When higher values of one variable are associated with lower values of the other (e.g. warmer weather is associated with lower heating bills) the relationship is said to be inverse or negative. Second, the scatter of points on the graph can be roughly summarized by a straight line, so we appear to have a linear relationship. Finally, the points are scattered around rather than upon the line, so the relationship is strong but not perfect: it is not possible to predict sales exactly when a given amount has been spent on advertising. Figure 6.13 illustrates other patterns which can occur on scatter diagrams. For the purpose of simulation we will divide these relationships into two categories: total dependence and diffuse dependence.

Simulating Total Dependence

Total dependence can be identified when all of the points in a scatter diagram fall on or extremely close to a line or a curve. If the relationship is positive it can be simulated simply by using the *same* random number to identify a value for both the independent and the dependent variable. For example, suppose that total production costs are totally dependent upon output. If we generate a random number which gives an output which has only a 10% chance of being exceeded, then, by using the same random number, we would obtain a production cost which also has only a 10% of being exceeded.

If the relationship is inverse, and if we are using random numbers between 0 and 1, then we could use the original random number to generate a value for the independent variable, and 1.0 minus this random number to generate a value for the dependent variable.

Simulating Diffuse Dependence

When there is clearly a relationship between two variables, but where this relationship is not perfect, as in Figure 6.12, we can refer to the relationship as being diffuse. In practice, this is a much more common type of relationship than total dependence, and a number of methods have been proposed for simulating it. One approach (see Hertz[12]) is referred to as *conditional sampling*. This involves the elicitation of a series of probability distributions for the dependent variable with each distribution being elicited on the assumption that a particular value of the independent variable has occurred. For example, we might ask the decision maker to estimate a probability distribution for annual delivery costs given that a certain sales level will be achieved. We would then repeat the process for other possible sales levels. The simulation process would first generate a sales level. This sales level would then determine which probability distribution should be used to generate a value for annual delivery costs. The obvious problem with this approach is that it will involve the decision maker in the assessment of a large number of probability distributions if the independent variable can take on one of many possible values.

Clearly, the approach can be simplified if we can assume that the basic *shape* of the dependent variable's probability distribution is the same, whatever the value of the independent variable. The PREDICT! computer package (see next section) allows the user to create a scatter diagram and then to input two boundary lines which show the maximum and minimum

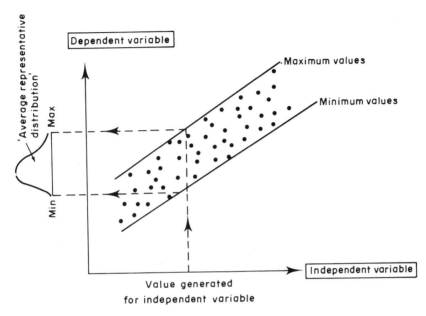

Figure 6.14 The method used by the PREDICT! package to simulate diffuse dependence

possible values of the dependent variable for different values of the independent variable (see Figure 6.14). The package then determines from the scatter diagram an 'average representative distribution' to describe the pattern of points between the two lines. When a value for the independent variable is generated, the package identifies the corresponding minimum and maximum values of the dependent variable, and then uses its average representative distribution to generate a value for the dependent variable which falls between the two boundary values.

The @RISK (pronounced 'at risk') computer package (also see next section) models dependence by asking the user either to input an equation to describe the nature of the relationship or a 'dependency function'. This dependency function requires the specification of a correlation coefficient to measure the strength of the association between the two variables. For other approaches to modeling dependence between variables see Eilon and Fawkes[13] and Hull.[14]

SUMMARY

In this chapter we have shown that simulation can be a powerful tool when the direct calculation of probabilities in a decision problem would be extremely complex. Moreover, because the approach does not involve advanced mathematics the methods involved and the results of analyses are accessible to decision makers who are not trained mathematicians or statisticians. Obviously, the application of simulation requires the use of a computer and a number of specialist computer packages are available. Two examples are @RISK, a Lotus 1-2-3 add-in (developed by the Palisade Corporation, Newfield, New York) and PREDICT! (Risk Decisions/Unison Technology Inc., Coraopolis, PA) which is free standing but has a spread-sheet-type display.

While computers excel at the rapid calculation of a large number of simulation results, we should not forget that the judgments of the decision maker are still the crucial inputs to the decision model. This means that, as always, care should be taken in the design of the model and in the elicitation of probabilities and utilities. The process of initially dividing the problem into small parts should help the decision maker in this task. Nevertheless, it is important to be aware of any assumptions and simplifications which are inherent in the model which is being used.

EXERCISES

(1) A small retail store sells a particular brand of monochrome and color television sets. Each monochrome set that is sold earns a profit of $100, while each color set earns $200 profit. The manager of the store estimates that the weekly demands for each type of set follow the probability distributions shown below. It can be assumed that the demands for each type of set are independent, as is the week-to-week demand.

Demand per week	Probability	
	Monochrome sets	Color sets
0	0.2	0.4
1	0.5	0.6
2	0.3	

(a) Determine the possible total profits that can be earned in any given week by selling television sets and *calculate* the probability of each of these profits being earned.

(b) The following two-digit random numbers have been generated by a computer. Use these numbers to simulate the demand for the two types of set for a 10-week period and hence calculate the profit that will be earned in each week. (The first set of numbers should be used for monochrome sets and the second for color.)

| Monochrome | 71 | 82 | 19 | 50 | 67 | 29 | 95 | 48 | 84 | 32 |
| Color | 36 | 44 | 64 | 92 | 39 | 21 | 18 | 55 | 77 | 73 |

(c) Use your simulation results to estimate the probability of particular profits being earned in a given week. How close are these probabilities to those that you calculated in (a)?

(2) The managers of a food company are about to install a number of automatic vending machines at various locations in a major city. A number of types of machine are available and the managers would like to choose the design which will minimize the profit that will be lost because the machine is out of order. The following model is to be used to represent the lost profit:

Cost of lost profit per month
= (number of breakdowns per month)
× (time to repair machine after each breakdown, in hours)
× (profit lost per hour)

One machine that is being considered is the Supervend, and the following probability distributions have been estimated for this machine:

Number of breakdowns per month	Prob.	Repair time (hours)	Prob.	Average profit lost per hour	Prob.
0	0.5	1	0.45	$10	0.7
1	0.3	2	0.55	$20	0.3
2	0.2				

(a) Use a table of random numbers, or the random number button on a calculator, to simulate the operation of a machine for 12 months and hence estimate a probability distribution for the profit that would be lost per month if the machine was purchased.

(b) Explain why the model is likely to be a simplification of the real problem.

(3) Trentware plc is a medium-sized pottery manufacturer which is based in the English Potteries. The company has fared badly over the last 10 years mainly as a result of Japanese competition, and recently this has led to major changes at the senior management level. The new Managing Director is determined to increase the company's market share, and you are a member of the ambitious new management team which intends to extend the company's range of tableware designs. Trentware's immediate objective is to launch a new high-quality product for the Christmas market in 18 months' time. Over 30 possible designs have been subjected to both technical analysis (to assess their production

implications) and extensive market research. As a result of this analysis, the number of potential designs has now been winnowed down to six. Some of these designs are thought to offer more risk than others because of changes in fashion, similarities to competing products and possible production problems. Now one design has to be selected from the six remaining. Risk analysis has been applied to each of these six designs and some of the main results are given below:

Design	1	2	3	4	5	6
Mean NPV (£0000)	50	21	20	46	-49	60
Standard deviation of NPV (£0000)	3	2	29	8	31	30

(a) You have been asked to explain to a colleague, who is unfamiliar with risk analysis, how these results are likely to have been derived. Draft some notes for your colleague and include in your notes an evaluation of the usefulness of the technique.

(b) Compare the risk analysis results for the six designs and discuss how a decision could be made between them.

(4) (This exercise is really designed to be carried out by a group of people, with each individual using a different set of random numbers. The individual results can then be combined by using the table at the end of the question.)

An equipment hire company has to decide whether to buy a specialized piece of earth-digging machinery for $6000. The machine would be sold after two years. The main factors which it is thought will affect the return on the investment are:

(i) The revenue generated by hiring the machine out for a day: it is certain that this will be $40;

(ii) The number of days that the machine will be hired out in year 1 and in year 2;

(iii) The costs of having the machine available for hire (e.g. maintenance and repair costs) in year 1 and year 2;

(iv) The price which will be obtained for the machine when it is sold at the end of year 2.

For each factor the following probability distributions have been estimated:

No of days hired out in year 1	Prob. (%)	No. of days hired out in year 1
under 100	30	(This is assumed to
100 to under 200	50	have the same
200 to under 300	20	distribution as year 1)

Annual costs ($)	Prob. in year 1 (%)	Prob in year 2 (%)
1000 to under 2000	50	30
2000 to under 3000	30	40
3000 to under 4000	20	30

Selling price ($)	Prob. (%)
1000 to under 2000	40
2000 to under 3000	60

Carry out one simulation of a possible combination of circumstances and calculate the NPV for your simulation Use a discount rate of 10%.

The results for the *entire group* can then be entered into the following table:

Net present value ($)	No. of simulations resulting in NPVs in this range	Probability
−15 000 to under 0	. . . .	. . . .
0 to under 5 000	. . . .	. . . .
5 000 to under 10 000	. . . .	. . . .
10 000 to under 15 000	. . . .	. . . .

Therefore the most likely range for the NPV appears to be . . . . and the probability of a negative NPV appears to be . . . .

(5) The managers of a chemical company have to decide whether to extend their existing plant or replace it with completely new equipment. A simulation of the two alternatives gives the following probability distributions of net present value:

NPV ($m)	Probabilities	
	Extend existing plant	Replace with new equipment
−3 to under −2	0.05	0.00
−2 to under −1	0.05	0.05
−1 to under 0	0.15	0.15
0 to under 1	0.29	0.26
1 to under 2	0.22	0.21
2 to under 3	0.14	0.18
3 to under 4	0.10	0.10
4 to under 5	0.00	0.05

(a) Compare the two distributions, and stating any necessary assumptions, determine the option that the management should choose.

(b) After the above simulations have been carried out a third possible course of action becomes available. This would involve the movement of some of the company's operations to a new site. A simulation of this option generated the following probability distribution. Is this option worth considering?

NPV ($m)	Probability
−2 to under −1	0.05
−1 to under 0	0.05
0 to under 1	0.40
1 to under 2	0.40
2 to under 3	0.04
3 to under 4	0.03
4 to under 5	0.03

The Standard Deviation

The standard deviation is designed to measure the amount of variation in a set of data: the larger the standard deviation, the more variation there is. Its calculation involves the following steps:

1. Calculate the mean of the observations.
2. Subtract the mean from each observation. The resulting values are known as deviations.
3. Square the deviations.
4. Sum the squared deviations.
5. Divide this sum by the number of observations.
6. Find the square root of the figure derived in (5).

EXAMPLE

The profits resulting from five products are shown below. Find the standard deviation.

$18\,000, \$19\,000 \$19\,000 \$21\,000 \$18\,000

The mean profit is $19\,000, so we have:

Profit ($)	Profit − mean	(Profit − mean)^2
18 000	− 1000	1 million
19 000	0	0
19 000	0	0
21 000	2000	4 million
18 000	− 1000	1 million
	Sum	6 million

$$\text{Therefore the standard deviation} = \sqrt{\frac{6 \text{ million}}{5}} = \$1095$$

The square of the standard deviation is known as the variance.

REFERENCES

1. Hertz, D. B. and Thomas, H. (1983) *Risk Analysis and its Applications*, Wiley, Chichester.
2. Singh, G. and Kiangi, G. (1987) *Risk and Reliability Appraisal on Micro Computers*, Chartwell-Bratt, Lund.
3. For a theoretical treatment of stochastic dominance see Hadar, J. and Russell, W. R. (1969) Rules for Ordering Uncertain Prospects, *The American Economic Review* March, 25–34.
4. Markowitz, H. M. (1959) *Portfolio Selection*, Wiley, New York.
5. Weston, J. F. and Brigham, E. F. (1978) *Managerial Finance* (6th edn), Holt, Rinehart and Winston, New York.
6. Drury, C. (1985). *Management and Cost Accounting*, Van Nostrand Reinhold, Wokingham.
7. Hespos, R. F. and Strassman, P. A. (1965) Stochastic Decision Trees in the Analysis of Investment decisions, *Management Science*, No. 10 B244–259.
8. Chapman, L. J. and Chapman, J. P. (1969) Illusory Correlation as an Obstacle to the Use of Valid Psychodiagnostic Signs, *Journal of Abnormal Psychology*, **74**, 271–280.
9. Tversky, A. and Kahneman, D. (1974) Judgement under Uncertainty: Heuristics and Biases, *Science*, **185**, 1124–1131.
10. Arkes, H. R., Harkness, A. R. and Biber, D. (1980) Salience and the Judgment of Contingency, *Paper presented at the Midwestern Psychological Association, St Louis, MO (36)* (as cited in H. R. Arkes and K. R. Hammond *Judgement and Decision Making*, (eds) 1986, Cambridge University Press, Cambridge).
11. Smedslund, J. (1963) The Concept of Correlation in Adults, *Scandinavian Journal of Psychology*, **4**, 165–173.
12. Hertz, D. B. (1979) Risk Analysis and Capital Investment, Harvard Business Review Classic, *Harvard Business Review*, September, 169–181.
13. Eilon, S. and Fawkes, T. R. (1973) Sampling Procedures for Risk Simulation, *Operational Research Quarterly*, **24**, No. 2, 241–252.
14. Hull, J. C. (1977) Dealing with Dependence in Risk Simulations, *Operational Research Quarterly*, **28**, No. 1, 201–213.

Chapter 7

Revising Judgments in the Light of New Information

INTRODUCTION

Suppose that you are a marketing manager working for an electronics company which is considering the launch of a new type of pocket calculator. On the basis of your knowledge of the market, you estimate that there is roughly an 80% probability that the sales of the calculator in its first year would exceed the break-even level. You then receive some new information from a market research survey. The results of this suggest that the sales of the calculator would be unlikely to reach the break-even level. How should you revise your probability estimate in the light of this new information?

You realize that the market research results are unlikely to be perfectly reliable: the sampling method used, the design of the questionnaire and even the way the questions were posed by the interviewers may all have contributed to a misleading result. Perhaps you feel so confident in your own knowledge of the market that you decide to ignore the market research and to leave your probability unchanged. Alternatively, you may acknowledge that because the market research has had a good track record in predicting the success of products, you should make a large adjustment to your initial estimate.

In this chapter we will look at the process of revising initial probability estimates in the light of new information. The focus of our discussion will be Bayes' theorem, which is named after an English clergyman, Thomas Bayes, whose ideas were published posthumously in 1763. Bayes' theorem will be used as a normative tool, telling us how we *should* revise our probability assessments when new information becomes available. Whether people actually do revise their judgments in the manner laid down by the theorem is an issue which we will discuss in Chapter 8.

Of course, new information, be it from market research, scientific tests or consultants, can be expensive to obtain. Towards the end of the chapter we will show how the potential benefits of information can be evaluated so that a decision can be made as to whether it is worth obtaining it in the first place.

BAYES' THEOREM

In Bayes' theorem an initial probability estimate is known as a *prior probability*. Thus the marketing manager's assessment that there was an 80% probability that sales of the calculator would reach break-even level was a prior probability. When Bayes' theorem is used to modify a prior probability in the light of new information the result is known as a *posterior probability*.

We will not put forward a mathematical proof of Bayes' theorem here. Instead, we will attempt to develop the idea intuitively and then show how a probability tree can be used to revise prior probabilities. Let us imagine that you are facing the following problem.

A batch of 1000 electronic components were produced last week at your factory and it was found, after extensive and time-consuming tests, that 30% of them were defective and 70% 'OK'. Unfortunately, the defective components were not separated from the others and all the components were subsequently mixed together in a large box. You now have to select a component from the box to meet an urgent order from a customer. What is the prior probability that the component you select is 'OK'? Clearly, in the absence of other information, the only sensible estimate is 0.7.

You then remember that it is possible to perform a 'quick and dirty' test on the component, though this test is not perfectly reliable. If the component is 'OK' then there is only an 80% chance it will pass the test and a 20% chance that it will wrongly fail. On the other hand, if the component is defective then there is a 10% chance that the test will wrongly indicate that it is 'OK' and a 90% chance that it will fail the test. Figure 7.1 shows these possible outcomes in the form of a tree. Note that because the test is better at giving a correct indication when the component is defective we say that it is *biased*.

When you perform the quick test the component fails. How should you revise your prior probability in the light of this result? Consider Figure 7.1 again. Imagine each of the 1000 components we start off with, traveling through one of the four routes of the tree. Seven hundred of them will follow the 'OK' route. When tested, 20% of these components (i.e. 140) would be expected to wrongly fail the test. Similarly, 300 components will follow the 'defective' route and of these 90% (i.e. 270) would be expected to fail the

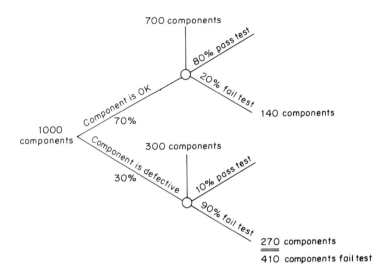

Figure 7.1 Tree diagram for the components problem

test. In total, we would expect 410 (i.e. 140 + 270) components to fail the test. Now the component you selected is one of these 410 components. Of these, only 140 are 'OK', so your posterior probability that the component is 'OK' should be 140/410, which is 0.341, i.e.

$$p(\text{component OK}|\text{failed test}) = 140/410 = 0.341$$

Note that the test result, though it is not perfectly reliable, has caused you to revise your probability of the component being 'OK' from 0.7 down to 0.341.

Obviously, we will not wish to work from first principles for every problem we come across, and we therefore need to formalize the application of Bayes' theorem. The approach which we will adopt, which is based on probability trees, is very similar to the method we have just applied except that we will think solely in terms of probabilities rather than numbers of components.

Figure 7.2 shows the probability tree for the components problem. Note that events for which we have prior probabilities are shown on the branches on the left of the tree. The branches to the right represent the new information and the conditional probabilities of obtaining this information under different circumstances. For example, the probability that a component will fail the test given that it is 'OK' is 0.2 (i.e. $p(\text{fails test}|\text{'OK'}) = 0.2$). Given that our component did fail the test, we are not interested in the branches labeled 'component passes test' and in future diagrams we will exclude irrelevant branches.

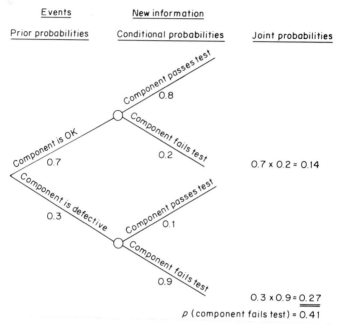

Figure 7.2 Applying Bayes' theorem to the components problem

We now calculate the probability of a component failing the test. First we determine the probability that a component will be 'OK' *and* will fail the test. This is, of course, a joint probability and can be found by applying the multiplication rule:

$$p(\text{OK and fails test}) = 0.7 \times 0.2 = 0.14$$

Next we determine the probability that a component will be defective *and* will fail the test:

$$p(\text{defective and fails test}) = 0.3 \times 0.9 = 0.27$$

Now a component can fail the test either if it is 'OK' or if it is defective, so we add the two joint probabilities to obtain:

$$p(\text{fails test}) = 0.14 + 0.27 = 0.41$$

The posterior probability is then found by dividing the appropriate joint probability by this figure. Since we want to determine the posterior probability that the component is 'OK' we select the joint probability which emanates from the 'component is OK' part of the tree, i.e. 0.14. Thus the posterior probability is 0.14/0.41, which is, of course, 0.341.

The steps in the process which we have just applied are summarized below:

1. Construct a tree with branches representing all the possible events which can occur and write the prior probabilities for these events on the branches
2. Extend the tree by attaching to each branch a new branch which represents the new information which you have obtained. On each branch write the conditional probability of obtaining this information given the circumstance represented by the preceding branch.
3. Obtain the joint probabilities by multiplying each prior probability by the conditional probability which follows it on the tree.
4. Sum the joint probabilities.
5. Divide the 'appropriate' joint probability by the sum of the joint probabilities to obtain the required posterior probability.

To see if you can apply Bayes' theorem, you may find it useful to attempt the following problem before proceeding. A worked solution follows the question.

Example

An engineer makes a cursory inspection of a piece of equipment and estimates that there is a 75% chance that it is running at peak efficiency. He then receives a report that the operating temperature of the machine is exceeding 80 °C. Past records of operating performance suggest that there is only a 0.3 probability of this temperature being exceeded when the machine is working at peak efficiency. The probability of the temperature being exceeded if the machine is not working at peak efficiency is 0.8. What should be the engineer's revised probability that the machine is operating at peak efficiency?

Answer

The probability tree for this problem is shown in Figure 7.3. It can be seen that the joint probabilities are:

$$p(\text{at peak efficiency and exceeds } 80\,^\circ\text{C})$$
$$= 0.75 \times 0.3 = 0.225$$

$$p(\text{not at peak efficiency and exceeds } 80\,^\circ\text{C})$$
$$= 0.25 \times 0.8 = 0.2$$

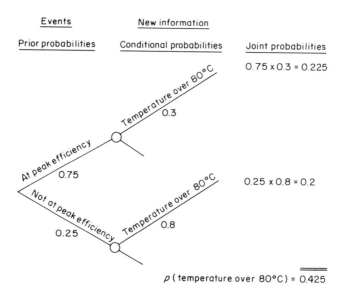

Figure 7.3 Applying Bayes' theorem to the equipment operating problem

so the sum of the joint probabilities is

$$0.225 + 0.2 = 0.425$$

and the required posterior probability is

$$0.225/0.425 = 0.529$$

Another Example

So far we have only applied Bayes' theorem to situations where there are just two possible events. The following example demonstrates that the method of handling a problem with more than two events is essentially the same.

A company's sales manager estimates that there is a 0.2 probability that sales in the coming year will be high, a 0.7 probability that they will be medium and 0.1 probability that they will be low. She then receives a sales forecast from her assistant and the forecast suggests that sales will be high. By examining the track record of the assistant's forecasts she is able to obtain the following probabilities:

p(high sales forecast given that the market will generate high sales)

$$= 0.9$$

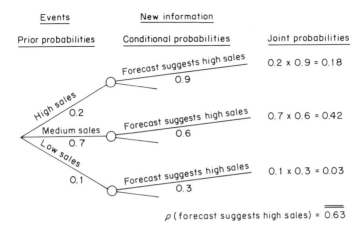

Figure 7.4 Applying Bayes' theorem to the sales manager's problem

p(high sales forecast given that the market will generate only medium sales)
$$=0.6$$

p(high sales forecast given that the market will generate only low sales)
$$=0.3$$

What should be the sales manager's revised estimates of the probability of (a) high sales, (b) medium sales and (c) low sales?

The tree for this problem is shown in Figure 7.4. The joint probabilities are:

p(high sales occur and high sales forecast) $=0.2\times0.9=0.18$
p(medium sales occur and high sales forecast)$=0.7\times0.6=0.42$
p(low sales occur and high sales forecast) $=0.1\times0.3=0.03$

So the sum of the joint probabilities is   0.63

which means that we obtain the following posterior probabilities:

p(high sales) $=0.18/0.63=0.2857$
p(medium sales)$=0.42/0.63=0.6667$
p(low sales) $=0.03/0.63=0.0476$

THE EFFECT OF NEW INFORMATION ON THE REVISION OF PROBABILITY JUDGMENTS

It is interesting to explore the relative influence which prior probabilities and new information have on the resulting posterior probabilities. Suppose that a geologist is involved in a search for new sources of natural gas in southern England. In one particular location he is asked to estimate, on the basis of a preliminary survey, the probability that gas will be found in that location. Having made his estimate, he will receive new information from a test drilling.

Let us first consider a situation where the geologist is not very confident about his prior probabilities and where the test drilling is very reliable. The 'vaguest' prior probability distribution that the geologist can put forward is to assign probabilities of 0.5 to the two events 'gas exists at the location' and 'gas does not exist at the location'. Any other distribution would imply that the geologist was confident enough to make some commitment in one direction. Clearly, if he went to the extreme of allocating a probability of 1 to one of the events this would imply that he was perfectly confident in his prediction. Suppose that having put forward the prior probabilities of 0.5 and 0.5, the result of the test driling is received. This indicates that gas is present and the result can be regarded as 95% reliable. By this we mean that there is only a 0.05 probability that it will give a misleading indication. (Note that we are assuming, for simplicity, that the test drilling is *unbiased*, i.e. it is just as likely to wrongly indicate gas when it is not there as it is to wrongly indicate the absence of gas when it is really present.) Figure 7.5 shows the probability tree and the calculation of the posterior probabilities. It can be seen that these probabilities are identical to the probabilities of the

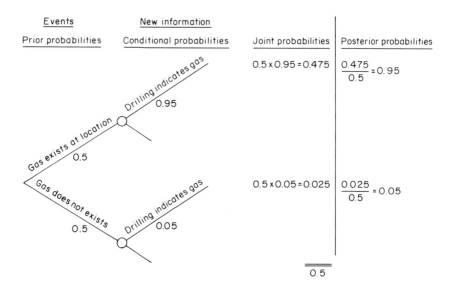

Figure 7.5 The effect of vague prior probabilities and very reliable information

test drilling giving a correct or misleading result. In other words, the posterior probabilities depend only upon the reliability of the new information. The 'vague' prior probabilities have had no influence on the result.

A more general view of the relationship between the 'vagueness' of the prior probabilities and the reliability of the new information can be seen in Figure 7.6. In the figure the horizontal axis shows the prior probability that gas will be found, while the vertical axis represents the posterior probability when the test drilling has indicated that gas will be found. For example, if the prior probability is 0.4 and the result of the test drilling is 70% reliable then the posterior probability will be about 0.61.

The graph shows that if the test drilling has only a 50% probability of giving a correct result then its result will not be of any interest and the posterior

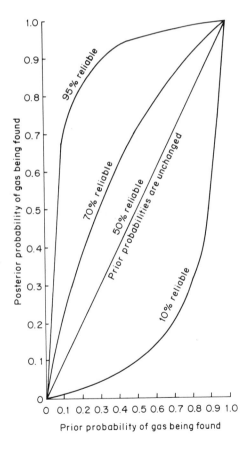

Figure 7.6 The effect of the reliability of new information on the modification of prior probabilities for the gas-exploration problem

probability will equal the prior, as shown by the diagonal line on the graph. By considering the distance of the curves from the diagonal line, it can be seen that the more reliable the new information, the greater will be the modification of the prior probabilities. For any given level of reliability, however, this modification is relatively small either where the prior probability is high, implying that the geologist has a strong belief that gas will be found and the test drilling confirms his belief, or where the prior probability is very small, implying that he strongly believes that gas will not be found. In the latter case, so strong is his disbelief that he severely restricts the effect of the disconfirming evidence from the test drilling.

At the extreme, if your prior probability of an event occurring is zero then the posterior probability will also be zero. Whatever new information you receive, no matter how reliable it is, you will still refuse to accept that the event is possible. In general, assigning prior probabilities of zero or one is unwise. You may think that it is impossible that San Marino will become a nuclear power by the year 2000. However, if you hear that seismic instruments have detected signs of nuclear testing there then you should allow yourself some chance of being persuaded by this information that the event might just be possible. Assigning a very small but non-zero prior probability might therefore be wiser.

Ironically, if the new information has less than a 0.5 chance of being reliable its result is of interest and the more unreliable it is, the greater the effect it will have on the prior probability. For example, if the test drilling is certain to give the wrong indication then you can be sure that the opposite of what has been indicated is the true situation!

APPLYING BAYES' THEOREM TO A DECISION PROBLEM

We will now consider the application of Bayes' theorem to a decision problem: a process which is sometimes referred to as *posterior analysis*. This simply involves the use of the posterior probabilities, rather than the prior probabilities, in the decision model.

A retailer has to decide whether to hold a large or a small stock of a product for the coming summer season. A payoff table for the courses of action and outcomes is shown below:

(Profits)

Decision	Low sales	High sales
Hold small stocks	$80 000	$140 000
Hold large stocks	$20 000	$220 000

The following table shows the retailer's utilities for the above sums of money (it can be assumed that money is the only attribute which he is concerned about):

Profit	$20 000	$80 000	$140 000	$220 000
Utility	0	0.5	0.8	1.0

The retailer estimates that there is a 0.4 probability that sales will be low and a 0.6 probability that they will be high. What level of stocks should he hold?

A decision tree for the retailer's problem in shown in Figure 7.7(a). It can be seen that his expected utility is maximized if he decides to hold a small stock of the commodity.

Before implementing his decision the retailer receives a sales forecast which suggests that sales will be high. In the past when sales turned out to be high the forecast had correctly predicted high sales on 75% of occasions. However, in seasons when sales turned out to be low the forecast had wrongly predicted high sales on 20% of occasions. The underlying market conditions are thought to be stable enough for these results to provide an accurate guide to the reliability of the latest forecast. Should the retailer change his mind in the light of the forecast?

We can uses Bayes' theorem to modify the retailer's prior probabilities in the light of the new information. Figure 7.7(b) shows the probability tree and the appropriate calculations. It can be seen that the posterior probabilities of low and high sales are 0.15 and 0.85, respectively. These probabilities replace the prior probabilities in the decision tree, as shown in Figure 7.7(c). It can be seen that holding a large stock would now lead to the highest expected utility, so the retailer should change is mind in the light of the sales forecast.

ASSESSING THE VALUE OF NEW INFORMATION

New information can remove or reduce the uncertainty involved in a decision and thereby increase the expected payoff. For example, if the retailer in the previous section was, by some means, able to obtain perfectly accurate information about the summer demand then he could ensure that his stock levels exactly matched the level of demand. This would clearly lead to an increase in his expected profit. However, in many circumstances it may be expensive to obtain information since it might involve, for example, the use of scientific tests, the engagement of the services of a consultant or the need to carry out a market research survey. If this is the case, then the question arises as to whether it is worth obtaining the information in the first place or,

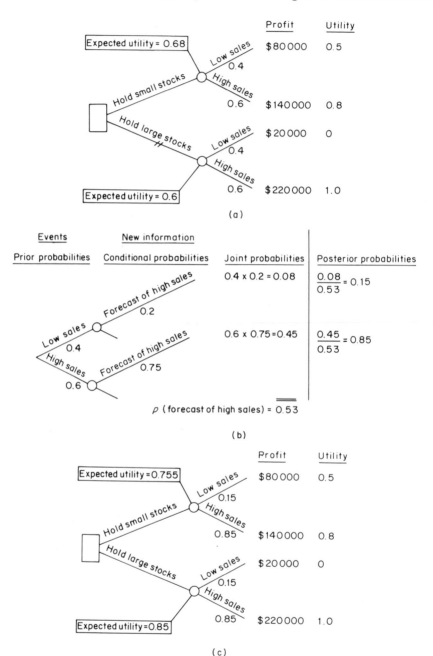

Figure 7.7 (a) A decision tree for the retailer's problem based on prior probabilities; (b) applying Bayes' theorem to the retailer's problem; (c) a decision tree for the retailer's problem using posterior probabilities

if there are several potential sources of information, which one is to be preferred (sometimes the process of determining whether it is worth obtaining new information is referred to as *preposterior analysis*). To show how this question can be answered, we will first consider the case where the information is perfectly reliable. (i.e. it is certain to give a correct indication) and then look at the much more common situation where the reliability of the information is imperfect.

The Expected Value of Perfect Information

In many decision situations it is not possible to obtain perfectly reliable information, but nevertheless the concept of the expected value of perfect information (EVPI) can still be useful. It might enable a decision maker to say, for example: 'It is unlikely that this consultant's predictions of our sales will be perfectly accurate, but even if they were, he would only increase my expected returns by $10 000. Since he is asking a fee of $15 000 it is clearly not worth engaging him.'

We will use the following problem to show how the value of perfect information can be measured. For simplicity, we will assume that the decision maker is neutral to risk so that the expected monetary value criterion can be applied.

A year ago a major potato producer suffered serious losses when a virus affected the crop at the company's North Holt farm. Since then, steps have been taken to eradicate the virus from the soil and the specialist who directed these operations estimates, on the basis of preliminary evidence, that there is a 70% chance that the eradication programme has been successful.

The manager of the farm now has to decide on his policy for the coming season and he has identified two options:

(1) He could go ahead and plant a full crop of potatoes. If the virus is still present an estimated net loss of $20 000 will be incurred. However, if the virus is absent, an estimated net return of $90 000 will be earned.
(2) He could avoid planting potatoes at all and turn the entire acreage over to the alternative crop. This would almost certainly lead to net returns of $30 000.

The manager is now informed that Ceres Laboratories could carry out a test on the farm which will indicate whether or not the virus is still present in the soil. The manager has no idea as to how accurate the indication will be or the fee which Ceres will charge. However, he decides initially to work on the assumption that the test is perfectly accurate. If this is the case, what is the maximum amount that it would be worth paying Ceres to carry out the test?

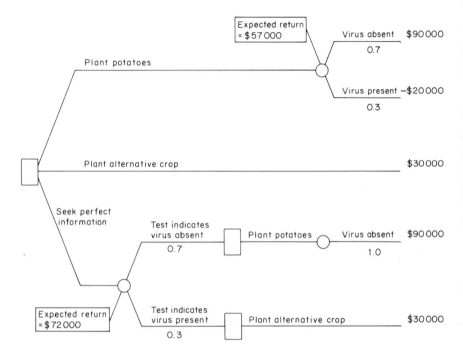

Figure 7.8 Determining the expected value of perfect information

A decision tree for the farm manager's problem is shown in Figure 7.8. In the absence of information from the test, he should plant a full crop of potatoes since his expected return if he follows this course of action will be:

$$0.7 \times \$90\,000 + 0.3 \times -\$20\,000 = \$57\,000$$

which exceeds the $30 000 return he will receive if he plants an alternative crop.

Now we need to determine the expected value of the perfect information which will be derived from the test. To do this, we need to consider each of the possible indications the test can give, how probable these indications are and how the manager should respond to each indication.

The calculations are summarized in Table 7.1. First, the test might indicate that the virus is absent from the soil. The specialist has said that there is a 70% chance that the virus is absent so, because the test is assumed to be perfectly accurate, the manager can assume that there is a probability of 0.7 that the test will give this indication. If the test does indicate that the virus is absent then the manager would earn $90 000 by planting potatoes and $30 000 planting the alternative crop so the best decision would be to plant potatoes.

Alternatively, the test might indicate that the virus is still present. There is a 0.3 probability that it will give this indication. In this case, the manager would lose $20 000 if he planted potatoes, so the best decision would be to plant the alternative crop and earn a net return of $30 000.

To summarize: there is a 0.7 probability that the test will indicate that the virus is absent, in which case the manager would earn net returns of $90 000, and a 0.3 probability that it will indicate that the virus is present, in which case he will earn $30 000. This means that his expected returns if he buys the perfect information will be $72 000.

Table 7.1 Calculating the expected value of perfect information

Test indication	Prob.	Best course of action	Payoff ($)	Prob. × payoff ($)
Virus is absent	0.7	Plant potatoes	90 000	63 000
Virus is present	0.3	Plant alternative	30 000	9 000
Expected payoff *with* perfect information =				72 000
Best expected payoff without perfect information =				57 000
Expected value of perfect information (EVPI) =				15 000

As we saw earlier, without the test the manager should plant potatoes when he would earn an expected return of $57 000. So the expected increase in his returns resulting from the perfect information (i.e. the expected value of perfect information) would be $72 000 – $57 000, which equals $15 000. Of course, we have not yet considered the fee which Ceres would charge. However, we now know that if their test is perfectly accurate it would not be worth paying them more than $15 000. It is likely that their test will be less than perfect, in which case the information it yields will be of less value. Nevertheless, the EVPI can be very useful in giving an upper bound to the value of new information. We next consider how to calculate the value of information which is not perfectly reliable.

The Expected Value of Imperfect Information

Suppose that, after making further enquiries, the farm manager discovers that the Ceres test is not perfectly reliable. If the virus is still present in the soil the test has only a 90% chance of detecting it, while if the virus has been eliminated there is a 20% chance that the test will incorrectly indicate its presence. How much would it now be worth paying for the test? To answer this question it is necessary to determine the expected value of imperfect information (EVII). As with the expected value of perfect information, we will need to consider the possible indications the test will give, what the probabilities of these indications are and the decision the manager should take in the light of a given indication.

The new decision tree for his problem is shown in Figure 7.9. If the manager decides not to buy the test then the decision is the same as before: he should plant potatoes, because the expected return on this option is $57 000. If he decides to buy the test then he will obviously wait for the test result before making the decision. The values missing from Figure 7.9, and represented by question marks, are the probabilities of the test giving each of the two indications and the probabilities of the virus being present or absent in the light of each indication.

Let us first consider the situation where the test indicates that the virus is present. We first need to calculate the probability of the test giving this indication using the probability tree in Figure 7.10(a). The prior probability of the virus being present and the test correctly indicating this is 0.3×0.9, which is 0.27. Similarly, the probability of the virus being absent and the test incorrectly indicating its presence is 0.7×0.2, which is 0.14. This means that the total probability that the test will indicate that the virus is present is $0.27 + 0.14$, which is 0.41. We can now put this probability onto the decision tree (Figure 7.11).

We can also use the probability tree in Figure 7.10(a) to calculate the posterior probabilities of the virus being present or absent if the test gives an

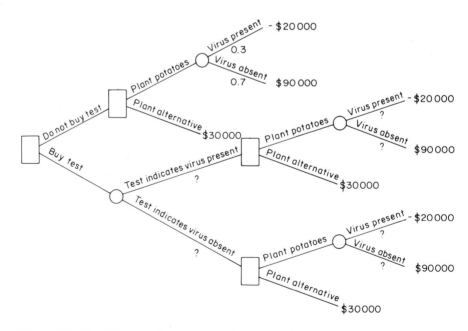

Figure 7.9 Deciding whether or not to buy imperfect information

indication of its presence. Using Bayes' theorem, it can be seen that these probabilities are 0.66 and 0.34, respectively. These probabilities can also be added to the decision tree.

We next consider the situation where the test indicates that the virus has been eliminated. Figure 7.10(b) shows the appropriate calculations. The probability of the test giving this indication is 0.59 (we knew that it would be 1−0.41 anyway) and the posterior probabilities of the presence and absence of the virus are 0.05 and 0.95, respectively. Again, these probabilities can now be added to the decision tree.

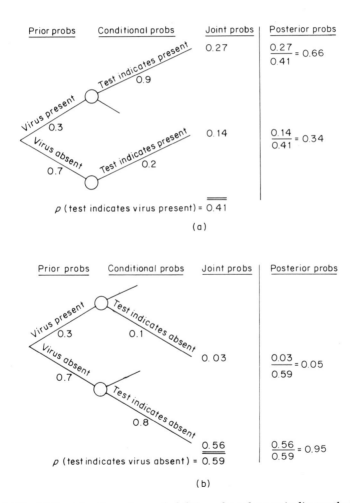

Figure 7.10 (a) Revising the prior probabilities when the test indicates that the virus is present; (b) revising the prior probabilities when the test indicates that the virus is absent

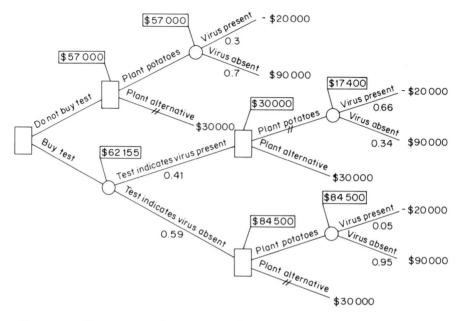

Figure 7.11 Determining the expected value of imperfect information

Let us now determine the expected payoff of buying the test by rolling back this part of the tree. If the test indicates that the virus is present then the best decision is to plant the alternative crop and earn $30 000. However, if the test indicates the absence of the virus, then clearly the best decision is to plant the potato crop, since the expected payoff of this course of action is $84 500. This means that if the manager buys the test there is a 0.41 probability that it will indicate the presence of the virus, in which case the payoff will be $30 000, and a 0.59 probability that it will indicate the absence of the virus, in which case the expected payoff is $84 500. So we have:

the expected payoff *with* the imperfect information
  from the test = 0.41 × $30 000 + 0.59 × 84 500   = $62 155
the expected payoff without the test   = $57 000

so the expected value of the imperfect information   = $5 155

It would not, therefore, be worth paying Ceres more than $5155 for the test. You will recall that the expected value of perfect information was $15 000, so the value of information from this test is much less than that from a perfectly reliable test. Of course, the more reliable the new information, the closer its expected value will be to the EVPI.

A summary of the main stages in the above analysis is given below:

(1) Determine the course of action which would be chosen using only the prior probabilities and record the expected payoff of this course of action;
(2) Identify the possible indications which the new information can give;
(3) For each indication:
    (a) Determine the probability that this indication will occur;
    (b) Use Bayes' theorem to revise the probabilities in the light of this indication;
    (c) Determine the best course of action in the light of this indication (i.e. using the posterior probabilities) and the expected payoff of this course of action;
(4) Multiply the probability of each indication occurring by the expected payoff of the course of action which should be taken if that indication occurs and sum the resulting products. This will give the expected payoff with imperfect information;
(5) The expected value of the imperfect information is equal to the expected payoff with imperfect information (derived in stage 4) less the expected payoff of the course of action which would be selected using the prior probabilities (which was derived in stage 1).

There is an alternative way of looking at the value of information. New information can be regarded as being of no value if you would still make the same decision regardless of what the information told you. If the farm manager were to go ahead and plant a crop of potatoes whatever the test indicated then there would clearly be no point in buying the test. Information has value if some of its indications would cause you to take a different decision than the one you would take without the benefit of the information. Let us calculate the expected value of the imperfect information derived from the test using this approach by again referring to Figure 7.11.

The decision based on the prior probabilities was to plant a crop of potatoes. If the test indicated the presence of the virus then you would make a different decision, that is, you would plant an alternative crop. Had you stayed with the prior decision, your expected payoff would have been $17 400, while planting the alternative crop yields $30 000. This means that the new information has stopped you from taking the inferior course of action and thereby increased your expected payoff by $12 600. However, if the test indicates that the virus is absent, you would make the same decision as you would without the information from the test, so the expected value of the information in this case is $0. Given the probabilities of the two test indications, the expected value of the imperfect information is:

$$0.41 \times \$12\,600 + 0.59 \times \$0 = \$5166$$

(The difference between this and the previous result is caused by rounding errors.)

As usual, the importance of sensitivity analysis cannot be understated, and in this respect spreadsheet packages are particularly useful. If the calculations involved in determining the EVII are carried out on a spreadsheet in the first place then it is relatively easy to examine the effect of variations in the inputs to the model. It would then be possible to make statements like the one which follows. 'I estimate that the proposed market research will be 90% reliable. Sensitivity analysis tells me that it would still be worth carrying out, even if its reliability was as low as 75% so I am fairly confident that we should go ahead and carry out the research.'

PRACTICAL CONSIDERATIONS

We will now outline a number of examples of the application of the methods we have just discussed and consider some of the practical problems involved. Clearly, it is easier to identify the expected value of perfect as opposed to imperfect information, and we recommend, that, in general, calculating the EVPI should be the first step in any information-evaluation exercise. The EVPI can act as a useful screen, since some sources of information may prove to be too expensive, even if they were to offer perfectly reliable data, which is unlikely.

Spetzler and Zamora[1] assessed the value of perfect information as part of the analysis of a problem faced by a large US corporation which had to decide whether or not to build a new production plant. If the corporation decided to build the plant it then had to decide whether to build an expansion to the plant at the same time as the original plant or whether to delay the expansion decision until the success of the original plant was known. The total expected value of the perfect information was estimated to be $11 million, which was high enough to suggest that the decision should be postponed until more information could be gathered. The value of having perfect information on particular aspects of the problem was also assessed. For example, one area of uncertainty was the amount of a valuable by-product which would be produced. The value of perfect information on this variable was estimated to be $6.2 million. In contrast, it was thought to be worth paying only $0.3 million to have perfect information on raw material costs. By comparing the value of perfect information for the different areas of uncertainty it was possible to identify those areas where the gathering of more information would be most useful.

As we saw in the previous section, assessing the expected value of imperfect information requires the decision maker to judge how reliable the information will be in order to obtain the conditional probabilities for the Bayes' theorem calculations. In some circumstances this assessment can be made on the basis of statistical theory. Consider, for example, a quality control problem where random sampling is being used to provide information on the proportion of a batch of components which are defective. It is possible to use elementary statistical methods to calculate the probability of a random sample of five components containing two defectives when, in reality, only 10% are defective. The track record of an information source can also be useful when estimating the conditional probabilities. For example, past records might suggest that on 20% of days when it rained the local weatherman had forecast fine weather. Similarly, experience with standard market tests leads to a common assumption that there is an 85% probability that they will give a correct indication (Scanlon[2]).

In most cases, however, the assessment of the reliability of the information will ultimately be based on the decision maker's subjective judgment. For example, Schell[3] used the expected value of imperfect information concept to assess the value of proposed information systems at a foods corporation, before a commitment was made to allocate personnel and other resources to the system. As part of the analysis, managers were asked to estimate the accuracy of proposed systems. Thus in the case of a sales forecasting system they would be asked to estimate the percentage of occasions that the system would predict a decrease in sales when sales would actually increase. The analysis acted as an initial screening process, leading to the early rejection of a number of projects and enabling the corporation to focus only on those which had a good potential for generating economic gains. Because the managerial and other inputs to the model were 'soft', Schell stressed the importance of sensitivity analysis. This involved the calculation of the expected value of perfect information and also the variation of the inputs into the model. He found that use of the Lotus 1-2-3 spreadsheet package made this sensitivity analysis easy, allowing a large number of 'what if' situations to be analyzed in minutes.

One of the areas with the greatest potential for the application of Bayesian analysis is market research. Indeed, Assmus[4] has argued that 'no other method has demonstrated an equally strong potential for analyzing the returns from marketing research'. Nevertheless, Lacava and Tull[5] refer to evidence that the approach is being used by only a small proportion of companies engaged in market research because of perceived difficulties in providing the necessary judgmental inputs and unfamiliarity with the calculations required to apply the technique. The authors have therefore developed a modified procedure for assessing the expected value of market research in decisions where a company has to decide whether or not to introduce a new product. The inputs required by the decision maker are: (1) the maximum loss which will need to be incurred before the product is removed from the market, (2) the probability of incurring this loss if the product is introduced, (3) the probability that the product will be successful, if introduced and (4) the probability that the market research will accurately predict the true state of the market. The authors have produced sets of tables which enable the expected value of the market research information to be determined, thus obviating the need to carry out calculations.

In the tables the EVII is expressed as the maximum percentage of the potential loss which should be spent on research. For example, suppose that the maximum loss is $1.2 million, the probability that this sum will be lost is 0.3, the probability that the product will be a success is 0.6 and the probability that the market research will indicate correctly that the product should be introduced is 90%, then the tables reveal that the expected value of the market research information is 11.57% of the maximum loss, that is, 11.57% of $1.2 million, which is $138 840.

SUMMARY

In this chapter we have discussed the role that new information can play in revising the judgments of a decision maker. We argued that Bayes' theorem shows the decision maker how his or her judgments should be modified in the light of new information, and we showed that this revision will depend both upon the 'vagueness' of the prior judgment and the reliability of the new information. Of course, receiving information is often a sequential process. Your prior probability will reflect the information you have received up to the point in time when you make your initial probability assessment. As each new instalment of information arrives, you may continue to revise your probability. The posterior probability you had at the end of last week may be treated as the prior probability this week, and be revised in the light of this week's information.

We also looked at how the value of new information can be assessed. The expected value of perfect information was shown to be a useful measure of the maximum amount that it would be worth paying for information. Calculating the expected value of imperfect information was seen to be a more involved process, because the decision maker also has to judge the reliability of the information. Because of this, we stressed the importance of sensitivity analysis, which allows the decision maker to study the effect of changes in these assessments.

EXERCISES

(1) The sales of a magazine vary randomly: in 70% of weeks they are classified as being 'high' while in 30% of weeks they are classified as 'low'.

(i) Write down prior probabilities of high and low sales in the coming week . . . .

(ii) You are now given the sales figures for Monday and these show low sales. In the past:

In weeks when sales turned out to be high, Monday had low sales on only 20% of occasions;
In weeks when sales turned out to be low, Monday had low sales on 60% of occasions.

Revise your prior probabilities in the light of Monday's sales figures.

(2) In January a sales manager estimates that there is only a '30% chance' that the sales of a new product will exceed one million units in the coming year. However, he is then handed the results of a sales forecast. This suggests that the sales will exceed one million units. The probability that this indication will be given when sales will exceed a million units is 0.8. However, the probability that the forecast will give this indication when sales will not exceed a million units is 0.4. Revise the sales manager's estimate in the light of the sales forecast.

(3) The probability of a machine being accidentally overfilled on a given day is 0.05. If the machine is overfilled there is a 0.8 probability that it will break down during the course of the day. If the machine is not overfilled the probability of a breakdown during the day is only 0.1. Yesterday the machine broke down. What is the probability that it had been accidentally overfilled?

(4) A mining company is carrying out a survey in a region of Western Australia. On the basis of preliminary results, the company's senior geologist estimates that there is a 60% probability that a particular mineral will be found in quantities that would justify commercial investment in the region. Further research is then carried out and this suggests that commercially viable quantities of the mineral will be found. It is estimated that this research has a 75% chance of giving a correct indication. Revise the senior geologist's prior probability in the light of the research results.

(5) A company which manufactures compact discs has found that demand for its product has been increasing rapidly over the last 12 months. A decision now has to be made as to how production capacity can be expanded to meet this demand. Three alternatives are available:

(i) Expand the existing plant;

(ii) Build a new plant in an industrial development area;

(iii) Sub-contract the extra work to another manufacturer.

The returns which would be generated by each alternative over the next 5 years have been estimated using three possible scenarios:

(i) Demand rising at a faster rate than the current rate;
(ii) Demand continuing to rise at the current rate;
(iii) Demand increasing at a slower rate or falling.

These estimated returns, which are expressed in terms of net present value, are shown below (net present values in $000s):

Course of action	Demand rising faster	Scenario Demand rising at current rate	Demand increasing slowly or is falling
Expand	500	400	-150
Build new plant	700	200	-300
Subcontract	200	150	-50

(a) The company's marketing manager estimates that there is a 60% chance that demand will rise faster than the current rate, a 30% chance that it will continue to rise at the current rate and 10% chance that it will increase at a slower rate or fall. Assuming that the company's objective is to maximize expected net present value, determine
  (i) The course of action which it should take;
  (ii) The expected value of perfect information.

(b) Before the decision is made, the results of a long-term forecast become available. These suggest that demand will continue to rise at the present rate. Estimates of the reliability of this forecast are given below:

p(forecast predicts demand increasing at current rate when actual demand will rise at a faster rate) = 0.3

p(forecast predicts demand increasing at current rate when actual demand will continue to rise at the current rate) = 0.7

p(forecast predicts demand increasing at current rate when actual demand will rise at a slower rate or fall) = 0.4

Determine whether the company should, in the light of the forecast, change from the decision you advised in (a).

(c) Discuss the limitations of the analysis you have applied above and suggest ways in which these limitations could be overcome.

(6) The managers of a soft drinks company are planning their production strategy for next summer. The demand for their products is closely linked to the weather, and an analysis of weather records suggests the following probability distribution for the June to August period:

Weather conditions	Probability
Hot and dry	0.3
Mixed	0.5
Cold and wet	0.2
	1.0

The table below shows the estimated profits ($000s) which will accrue for the different production strategies and weather conditions:

	Weather conditions		
Production strategy	Hot and dry	Mixed	Cold and wet
Plan for high sales	400	100	– 100
Plan for medium sales	200	180	70
Plan for low sales	110	100	90

(a) On the basis of the information given, determine:
   (i) The course of action which will maximize expected profits;
   (ii) The expected value of perfect information
   and discuss the practical implications of your result.
(b) A long-range weather forecast suggests that next summer's weather conditions will, in general, be cold and wet. The reliability of the forecast is indicated by the following probabilities which are based on past performance:

p(cold, wet conditions forecast when weather will be hot and dry)
$$= 0.3$$

p(cold, wet conditions forecast when weather will be mixed)
$$= 0.4$$

p(cold, wet conditions forecast when weather will be cold and wet)
$$= 0.6$$

In the light of the long-range weather forecast, should the company change from the course of action you recommended in (a)?
(7) A company has just received some 'state of the art' electronic equipment from an overseas supplier. The packaging has been damaged during delivery and the company must decide whether to accept the equipment. If the equipment itself has not been damaged, it could be sold for a profit of $10 000. However, if the batch is accepted and it turns out to be damaged, a loss of – $5000 will be made. Rejection of the equipment will lead to no change in the company's profit. After a cursory inspection, the

company's engineer estimates that there is a 60% chance that the equipment has not been damaged.

The company has another option. The equipment could be tested by a local specialist company. Their test, however, is not perfectly reliable and has only an 80% chance of giving a correct indication.

How much would it be worth paying for the information from the test? (Assume that the company's objective is to maximize expected profit.)

(8) The managers of Red Valley Auto Products are considering the national launch of a new car-cleaning product. For simplicity, the potential average sales of the product during its lifetime are classified as being either high, medium or low and the net present value of the product under each of these conditions is estimated to be $80 million, $15 million and $-40 million, respectively. The company's marketing manager estimates that there is a 0.3 probability that average sales will be high, a 0.4 probability that they will be medium and a 0.3 probability that they will be low. It can be assumed that the company's objective is to maximize expected net present value.

(a) On the basis of the marketing manager's prior probabilities, determine:

(i) Whether the product should be launched;

(ii) The expected value of perfect information.

(b) The managers have another option. Rather than going immediately for a full national launch they could first test market the product in their Northern sales region. This would obviously delay the national launch, and this delay, together with other outlays associated with the test marketing, would lead to costs having a net present value of $3 million. The test marketing would give an indication as to the likely success of the national launch, and the reliability of each of the possible indications which could result are shown by the conditional probabilities in the table below (e.g. if the market for the product is such that high sales could be achieved there is a probability of 0.15 that test marketing would in fact indicate only medium sales):

| | Test marketing indication | | |
Actual national sales	High national sales	Medium national sales	Low national sales
High	0.80	0.15	0.05
Medium	0.25	0.60	0.15
Low	0.10	0.30	0.60

Calculate the expected value of imperfect information and hence determine whether the company should test market the product.

REFERENCES

1. Spetzler, C. S. and Zamora, R. M. (1984) Decision Analysis of a Facilities Investment and Expansion Problem in R. A. Howard and J. E. Matheson (eds) *The Principles and Applications of Decision Analysis*, Strategic Decision Group, Menlo Park, CA.
2. Scanlon, S. (1976) Test Marketing: Calling the Shots more Closely, *Portfolio Sales and Marketing Management*.
3. Schell, G. P. (1986) Establishing the Value of Information Systems, *Interfaces*, **16**:3, 82–89.
4. Assmus, G. (1977) Bayesian Analysis for the Evaluation of Marketing Research Expenditures, *Journal of Marketing Research*, **41**, 562–568.
5. Lacava, G. J. and Tull, D. S. (1982) Determining the Expected Value of Information for New Product Introduction, *Omega*, **10**, No. 4, 383–389.

Chapter 8

The Quality of Human Judgment: Laboratory Studies

INTRODUCTION

In earlier chapters we have shown how subjective probabilities can be utilized in decision models. A major question arises: how good are people at making these judgments? This chapter presents an overview of laboratory research that has been conducted on the human judgment of probability over the last 30 years. By 'laboratory research' we mean research conducted in relatively artificial situations, often with college students as participants in psychological experiments. As we shall see, such studies allow fine-grained analysis of human judgment but suffer from the potential criticism that their results have little to do with real-world decision making. The last 30 years has been chosen because at the beginning of this period laboratory research began in earnest, and over this period the view of human ability has changed from one of sub-optimality and bias to one that is now less sure of the capabilities of human judgment. The first section of the chapter reviews studies of the *overall* quality of human judgment. The second evaluates the worth of studying individual differences in decision making and judgment. Are some people better decision makers than others?

Interest in subjective probability judgments by psychologists can be dated to a seminal paper by Ward Edwards entitled 'The Theory of Decision Making' which appeared in the *Psychological Bulletin* in 1954. This article presented what was essentially an economic decision theory as a psychological or subjective decision theory. *Subjective expected utility* (SEU) *theory*, as it became known over the next few years, specified that normative decisions could be prescribed on the basis of two independent sources of information; subjective probabilities attached to the occurrence or non-occurrence of future events, and utilities or subjective values attached to the possible outcomes of the interplay between human actions and events at some time in the future.

As we have seen, in the 1960s this economic theory was incorporated into *decision analysis* and was pioneered as a way to improve decision making. However, business school-based proponents of decision analysis were more concerned with the computational aspects of the technology and less with the psychology of probability and value.

A major line of research that has been concerned with the quality of human judgment of probability has stemmed from the work of Tversky and Kahneman. In a series of well-written and accessible papers starting in the early 1970s, they outlined some of the heuristics, or rules of thumb, that people use for probability assessment.[1] Much of this research is now incorporated into general introductory texts, perhaps because their research studies are easily understood, non-technical and appealing to the non-specialist reader.

HEURISTICS IN PROBABILITY ASSESSMENT

To describe Tversky and Kahneman's approach to data collection we will present two questions which we would like you to consider.

1. Suppose I sample a word of three letters or more at random from an English text. Is it more likely that the word starts with an 'r' or that 'r' is its third letter?
2. Which cause of death is more likely out of each pair:
   (a) Lung cancer or stomach cancer;
   (b) Murder or suicide;
   (c) Diabetes or a motor vehicle accident?

If your responses are like those of most of Tversky and Kahneman's subjects you may have said that 'r' is more likely to start a word. However, in reality, 'r' is more frequent as the third letter. Tversky and Kahneman argue that people approach the problem by recalling words that begin with an 'r' (e.g. road) and words that have 'r' as the third letter (e.g. care). Because it is much easier to search for words by their first letter than by their third most people, Tversky and Kahneman argue, judge that words beginning with 'r' are more likely.

Tversky and Kahneman have also demonstrated that we judge the probability of an event by the ease with which relevant information about that event is imagined. Instances of frequent events are typically easier to recall than those of less frequent ones. Thus *availability* is often a valid cue for the assessment of frequency and probability. However, since availability is also influenced by factors unrelated to likelihood, such as familiarity, recency and emotional saliency, reliance on it *may* result in systematic biases. In a convincing study, Lichtenstein *et al.*[2] found that people overestimated the relative frequency of diseases or causes of death which are much publicized, such as murder or lung cancer, whereas the relative frequency of less-publicized causes of death, such as stomach cancer and diabetes, were underestimated. In all parts of question 2, above, the second alternative is about one and a half times more likely than the first. Most people incorrectly think that death by lung cancer is much more likely than death by stomach cancer and death by murder is more likely than suicide. Conversely, most people correctly think that death by motor vehicle accident is more likely than death caused by diabetes.

Another heuristic tending to bias probabilistic judgment is *representativeness*. Consider the following problem:

> This is a brief personality description of Tom W written by a psychologist when Tom was in his senior year at high school. Tom W is of high intelligence, although lacking in true creativity. He has a need for order and clarity and for neat and tidy systems in which every detail finds its appropriate place. His writing is rather dull and mechanical, occasionally enlivened by somewhat corny puns and by flashes of imagination of the sci-fi type. He has a strong drive for competence. He seems to have little feel and sympathy for other people and does not enjoy interacting with others. Self-centered, he nonetheless has a deep moral sense. This personality description has been chosen, at random, from those of 30 engineers and 70 social scientists. What is your probability that Tom W is an engineer?

You have probably answered that Tom W is more likely to be an engineer than a social scientist. However, Kahneman and Tversky argue that the base rate should have predominance over the low-reliability personality sketch, such that your probability response should be little different, if at all, from the base-rate probability of a 0.7 chance that Tom W is a social scientist. Using similar problems, Kahneman and Tversky found that when no individuating evidence is given, base rates are properly utilized, but when relatively worthless information is given, as in the above example, base rates are ignored. Kahneman and Tversky coined the term *representativeness* to refer to the dominance of individuating information in intuitive prediction.

Tversky and Kahneman have also identified a heuristic called *anchoring and adjustment*. To demonstrate this effect, subjects were asked to estimate various quantities, stated in percentages (e.g. the percentage of African countries in the United Nations). Before they made their estimates, the subjects were shown an arbitrary starting value between 0 and 100 given by the result of a spin of a wheel of fortune. Subjects were required to indicate whether they considered this value too high or too low and then to give their own estimate. Different groups of people were given different starting values. Surprisingly, the arbitrary starting values had a considerable influence on estimation. For example, median estimates for the question posed above were 25% and 45%, for groups which received 10% and 65% as starting points, respectively. Reward for accuracy did not reduce the anchoring and adjustment effect.

Slovic[3] gave another example of anchoring in the subjective valuation of gambles. He found that in making these judgments people who find a gamble basically attractive use the amount to win as a natural starting point. They then adjust the amount to win downward to take into account the less-than-perfect chance of winning and the possibility of losing a small amount. Typically, this adjustment is insufficient, and Slovic argued that this is why people price gambles inconsistently with straight choices between pairs of gambles where a monetary response is not required.

Yet another bias in judgment has been identified as *misperception of regression towards the mean*. Suppose a large number of children have been given an intelligence test. Some scored above the mean, some below. If an equivalent test was then administered to the children, those who scored above the mean in the first test would obtain a lower average score in the second, whereas those children who scored below the mean on the first test would obtain a higher average score in the second. Random fluctuations around the mean will produce sub-standard and above-standard performance which is highly likely to be followed by an improvement and a worsening, respectively. This phenomenon, known as *regression towards the mean*, was first documented by Galton a century ago. Failure to appreciate the importance of regression can have important consequences, as Tversky and Kahneman have documented. They found that flight instructors typically praised the trainee pilot after the successful execution of a flight maneuver and admonished him after a poor performance. Lack of understanding of regression towards the mean led the flight instructors to the erroneous conclusion that praise is detrimental to learning whereas punishment is beneficial. Tversky and Kahneman argue that people tend to use causal explanations to explain random variation.

Tversky and Kahneman have devised many simple paper and pencil tests that have revealed heuristics people commonly use when making judgments about uncertain events. According to Tversky and Kahneman, these heuristic principles may be quite useful in that they reduce the complexity of probability estimation, but they can lead to severe and systematic errors, analogous to the effects of the perceptual illusions. For example, perceptual psychology has documented that one cue to distance evaluation is the clarity of an object's image on the retina. The sharper the image of an object, the closer it appears to be. Clarity as a cue to distance has obvious general validity. However, when visibility is poor, distances are overestimated because the contours of an object are indistinct. Similarly, when visibility is good, distance is underestimated.

How can the biases of the heuristic principles be reduced so that the process of probability estimation is improved? In a paper entitled 'Intuitive Prediction: Biases and Corrective Procedures'[4] Kahneman and Tversky have attempted to answer this question. Their 'strategy of debiasing' essentially concentrates on the elicitation and recognition of the importance of the usually neglected, but otherwise available, base-rate information for use in prediction.

However, in the last five years the research of Tversky and Kahneman has become the subject of critical reaction. Their demonstrations of judgmental fallibility, it is argued, should not be taken as having pejorative implications for the rationality of human reasoning and may not always occur in circumstances more akin to real-world decision making.[5]

In one attack on the status of the 'most broadly accepted view that, in general, human judgment is seriously biased' Beach et al.[6] argued that this view is not supported by convincing data. First, Christensen-Szalanski and Beach identified what they termed the 'citation bias'. Of the 3500 articles on judgment and reasoning published between 1972 and 1981 only 84 were empirical studies. Of these, 47 obtained poor performance and 37 had good performance. However, poor performance results were cited an average of six times more often than were good performance results.

Second, Beach et al. argued that the 'word problems' developed by cognitive psychologists may not be fully understood by subjects and, further, may not generalize to workaday judgment and reasoning. For example, it has been demonstrated that very small changes in the experimenter's task can produce evidence of poor and good judgment. Pitz[7] gives an example of a situation in which poor probabilistic judgment is initially evident. He quotes a question previously utilized by Tversky and Kahneman:[8]

> The mean IQ of the population of eighth graders in a city is known to be 100. You have selected a random sample of 50 children for a study of educational achievements. The first child tested has an IQ of 150. What do you expect the mean IQ to be for the whole sample?

Tversky and Kahneman reported that a large number of subjects responded with an estimate of 100. They concluded that these subjects expected samples in the long run to generate extreme values in one direction that would cancel extreme values in the other direction, one version of the gambler's fallacy. Pitz noted, however, that it is possible that subjects are simply using a well-learned rule: 'Given a known population and a randomly selected sample from that population, the expected sample mean is equal to the population mean.' Pitz hypothesized that other rules may exist (for example, 'Given two random samples, the expected mean of the second sample is independent of information about the first sample'), but that they had a lower priority in that other responses or rules are likely to be used first. He demonstrated this by rewording Tversky and Kahneman's question in the following manner:

> The mean IQ of the population of eighth graders in a city is known to be 100. You have selected a random sample of 50 children for a study of educational achievements. The first child tested has an IQ of 150. What do you expect the mean IQ to be for the remaining 49 children?

As Pitz noted, people answering anything other than 100 are rare, and usually turn out to have misunderstood the question. From this finding, Pitz developed a theory of a 'production system' of decision rules within the individual that produces responses. A production system is a mechanism for selecting the particular rule to apply to a specific decision situation. In order to show the presence of a rule in a person's production system it is, of course, necessary to prevent decision rules being used that are more usually applied in the particular situation under study. Pitz's study provides an 'existence demonstration' of valid judgment showing that there are conditions in which good judgmental rules are used.

Beach *et al.* also make the point that people who typically take part in the word problem studies are undergraduate students who are unrepresentative of the people commonly regarded as qualified judges. In other words, the experimental subjects would not claim special *expertise* in the subject matter of the word problems. Certainly, completion of such word problems is not an everyday activity for most business decision makers! A final point that Beach *et al.* make is that judgment should not be isolated from action, since judgment is seldom an end in itself but a guide to subsequent action. Christensen-Szalanski has also questioned the generalizability of laboratory findings of cognitive biases in decision making to the real world, and concluded that, even if the biases are prevalent, this knowledge by itself is of little use since we need to 'determine the degree to which the decision outcome would improve once the cognitive bias was eliminated and whether this improvement merits the expenditure of resources needed to accomplish it'.[9]

Phillips[10] has argued that interest should now focus on *what people can do under favourable conditions* whereas, as we have seen, the research literature has tended to be dominated by reports of what people *actually do without help, guidance or training*. He makes the point that conditions need to be appropriate for the generation of precise reliable and accurate assessments of probability, and lists eight conditions which need to be satisfied. These include training in probabilistic thinking if the assessor is unfamiliar with probability concepts and the use of experts with substantive expertise in the area where judgments are required.

Generally, then, the findings of judgmental heuristics and biases in the psychological laboratory recently has been criticized with respect to the implications that can be drawn for the prevalence of judgmental fallibilities in real-world decision making. We shall return to this issue in the next section, where we describe and evaluate laboratory studies of dynamic probability revision.

In the mid-1960s Edwards was working with Bayes' theorem which, as we saw in Chapter 7, is a normative theory of the way in which subjective probabilities attached to the truth of hypotheses should be revised in the light of new information. Edwards was interested to see whether Bayes' theorem was descriptive of human revision of opinion. These studies are described next.

REVISION OF PROBABILISTIC OPINION

In a number of investigations, Edwards and his colleagues[11] found that unaided human revision of opinion often was *less* than the theorem would prescribe. In other words, posterior opinion, given updated information, was not as extreme as that calculated by Bayes' theorem. This result has been termed *conservatism*. Most of the laboratory research on the conservatism phenomenon used the 'book bag and poker chip' paradigm. The basic paradigm was as follows. The experimenter holds three opaque book bags. Each contained one hundred poker chips in different, but stated, proportions of red and blue. The experimenter chooses one bag at random, shuffles the poker chips inside and successively draws a single poker chip from the bag. After each draw, the poker chip is replaced and all the poker chips are shuffled before the next draw. The subjects' task is to say how confident (in probability terms) he or she is that the chosen bag is Bag 1, Bag 2 or Bag 3. The color of the poker chip drawn on each occasion from the same bag is information on which to revise prior probabilities of ⅓, ⅓ and ⅓ for the three bags.

How well does Bayes' theorem *describe* human opinion revision? The data from a large number of laboratory studies, using tasks very similar to the one we have described, show that the amount of opinion revision is often less than the theorem would prescribe. However, the *amount* of conservatism shown in a particular task is highly situation-specific.

The degree of conservatism has been shown to vary with the diagnosticity of the data. For example, imagine that you were being shown samples drawn from one of two opaque bags each containing a hundred colored balls. One of the bags contains 49 red balls and 51 blue balls, while the other contains 51 red and 49 blue. Clearly, two consecutive samples each of a blue ball would not be very diagnostic as to which bag was generating the data. Experiments have shown that the more diagnostic the data, the more conservative the experimental subject. When the data become very undiagnostic, as in the above example, human probability revision can become too extreme.

Another variable which affects the amount of conservatism exhibited is the way in which data are presented. According to Bayes' theorem, it should not matter whether a series of data is presented sequentially or all at once. However, other investigators[12] found that subjects' estimates were less conservative when a sample of data was presented one item at a time, with probability revisions required after each item, than when single estimates of the posterior probabilities were based on the information contained in the whole sample. A further finding was that, even when data are presented sequentially, the *ordering* or sequence of the data influences probability revision. Other researchers[13] described an 'inertia effect', where subjects tended *not* to revise their probabilities downward once the *initial* part of a sequence of data had favoured one of the hypotheses under consideration. In other words, subjects seemed unwilling to reduce their probabilities on favored hypotheses following disconfirming evidence.

A further set of researchers[14] investigated human probability revision in a situation they considered to be nearer to real life than the book bag and poker chip paradigm. They argued that the datum in the latter paradigm is usually restricted to one of two different types, e.g. a red or a blue poker chip, and that there are usually only two or three possible revisions that can be made on the information obtained. In real life, information may vary along a continuum rather than being discrete values. In their study, these investigators used a hypothesis set consisting of the populations of male and female heights. The subjects' task was to decide which population was being sampled on the basis of the data contained in randomly sampled heights from that population. Using this task, they found conservatism only *half* as great as with the typical book bag and poker chip task. They concluded that this effect was due to their subjects' greater familiarity with the data-generating process underlying their task.

Winkler and Murphy took the arguments one stage further. In an article entitled 'Experiments in the Laboratory and the Real World'[15] they argued that, even though the typical book bag and poker chip paradigm seems outwardly simple, it differs in four major respects from the real world.

First, the inference tasks so far discussed differ from everyday situations in that, in most of the laboratory tasks, samples of data are *conditionally independent*. That is, two or more pieces of information have an identical implication for the posterior opinion to be placed on a particular hypothesis no matter if the pieces of information are considered jointly or in *any* sequence. In the real world, Winkler and Murphy argue, successive items of information may, to a degree, be redundant. This would mean that the total impact of several pieces of information would be less than the sum of the impacts of each item observed separately. To quote Winkler and Murphy:

> Therefore, one possible explanation for conservatism in simple book bag and poker chip experiments is that the subject is behaving as he does in more familiar situations involving redundant information sources (p. 256).

Second, in most experiments using the book bag and poker chip paradigm the data generators (the book bags) are *stationary*. That is, the contents of the book bag remain the same during the experiment. In the real world our hypotheses may not remain constant. Indeed, the nature of information obtained may change our hypotheses.

Third, in the real world data may be *unreliable* and therefore be less diagnostic than perfectly reliable data, like the color of a poker chip. In many real-life opinion revisions the probability assessor not only has to determine the diagnosticity of a piece of data but also its reliability. In support of Winkler and Murphy's argument other investigators[16] have shown that, when laboratory tasks include unreliable data, human inference tends to be less conservative, but still not in accordance with Bayes' theorem.

Fourth, subjects in the book bag and poker chip experiment are typically given highly diagnostic data. In the real world, data may be relatively *undiagnostic* and so the result of subjects generalizing their experience of the real world to the novel laboratory task may result in conservatism.

In summary, Winkler and Murphy concluded that 'conservatism may be an artefact caused by dissimilarities between the laboratory and the real world'. However, despite these criticisms of the results of laboratory experiments, there has been a considerable research effort into the development of computer-aided systems that implement Bayes' theorem. In these systems the probability assessor makes the probability judgment after each item of information arrives, but the computer aggregates these assessments in an optimal manner. Of course, in contrast to the laboratory, it is usually impossible in real life to check the veracity of prior opinion, likelihoods and posterior opinion against a suitable agreed-upon criterion. This fact accounts for the rarity of studies of opinion revision in the real world that are analogous to those studied so intensively in the laboratory. Additionally, very few laboratory studies have appeared since the mid-1970s.

Generally, then, the study of the overall quality of human probability judgment in the psychological laboratory has been questioned. Do the results generalize outside contrived laboratory settings? In Chapter 9 we analyze the quality of judgment in contexts more akin to real-world forecasting practice. Next, however, we turn from looking at the overall quality of human judgment and evaluate the usefulness of studying individual differences in decision-making styles or abilities. The general theme is, are some people better at making decisions, or judgments, than others?

INDIVIDUAL AND SITUATIONAL INFLUENCES ON DECISION MAKING

This section presents a selective review of studies of individual and situational influences on decision making that have been reported in the psychological and management journals. Various decision situations have been studied, various correlates of decision making have been obtained and various decision styles have been proposed. The major difference between studies has been the relative emphasis on the decision maker or the decision situation as the main source of behavioral variation. As we shall see, such research has implications for the design of management decision support systems.

Personality and Decision Making

The research to be reviewed in this section has usually taken a 'personality' measure, a single decision situation and a sample of people and attempted to see if any relationship exists between personality and decision making. For instance, the literature on authoritarianism and dogmatism assumes that people who score high on scales measuring these concepts see the world in 'black and white' and make extreme judgments or responses. The primary characteristics of an individual who is intolerant of ambiguity are 'premature closure' and 'need for certainty'. For example, a negative response to the statement 'people who insist on a yes or no answer just don't know how complicated things really are' characterizes a person intolerant of ambiguity. Frenkel-Brunswik, writing in *The Authoritarian Personality*,[17] notes that 'a simple, firm, often stereotypical cognitive structure is required. There is no place for ambivalence or ambiguities. Every attempt is made to eliminate them.' Indeed, one investigator[18] has developed a measure of dogmatism based on the content analysis of 'qualifiers' in the publications of writers. For example, dogmatic writers would be expected to use qualifiers such as *always, never, nothing but, completely, must,* etc. whereas non-dogmatic writers would use qualifiers such as *often, rarely, greatly, considerably, can,* etc.

From these orthodox conceptualizations of the personality/cognitive measures Wright and Phillips[19] anticipated strong relationships with their own measures of probability assessment. These evaluate the tendency to use numeric probabilities or certainty assessments in response to questions concerning uncertain situations, and the realism or 'calibration' of the numeric assessments used. As we shall discuss more fully in Chapter 9, a person is said to be 'well calibrated' if the correct proportion is equal to the confidence or probability assigned. For instance, a well-calibrated probability assessor who said, in response to each of 10 questions, that he or she was 70% sure that he or she had chosen the correct answer would have selected that answer seven times. Similarly, all answers made with 100% certainty should turn out to be correct. Wright and Phillips found that high authoritarianism was related to overconfident use of 100% assessments and to a less fine discrimination in probability assessed numerically, which is shown by little usage of intermediate probabilities. Also, dogmatic individuals were less likely to say that they did not know the answer.

Other researchers[20] have investigated the relationship between characteristics of the decision maker and decision process using a decision simulation that requires a subject to play the role of a business manager who must select one of three hypothetical salesmen for promotion to sales manager. Various measures are taken during the simulation, including amount of information viewed, time required to reach a decision, and confidence in that decision. Using a battery of personality/cognitive measures, these researchers found that high dogmatism had only a moderate relationship to decisional confidence. Similarly, risk-taking propensity, as measured by another questionnaire,[21] had a very low correlation with time to take a decision. Other decision-maker attributes, including age, experience in making personnel decisions, and intelligence made only minor contributions in accounting for individual differences in decision making.

Another study[22] examined the relationship between scores on another battery of personality measures and three problems measuring risk-taking propensity. It was found that:

> Individuals exhibiting riskiness in decision-making were characterized as persistent, effective in their communication, confident and outgoing, clever and imaginative, aggressive, efficient and clear-thinking, and manipulative and opportunistic in dealing with others.

The last study to be described in this section[23] had subjects rate gambles in terms of hypothetical bids to play the bets. The investigator used two groups of subjects. The members of the first group were unfamiliar with statistics and were 'unable to give a proper definition or example of expected value', whereas the members of the second group, who were taking an introductory management course in which elementary statistics had been covered, were able to express understanding of expected value. It was found that the statistically trained subjects used more consistent decision rules in evaluating the gambles than the untrained ones. The statistically trained subjects tended to use expected value as an evaluation criterion and, in answer to protocol questions, also said they focused equally on all parts of the gamble compared to the untrained subjects. The untrained subjects, by contrast, tended to focus differentially on parts of the gambles such as the amount to win, probability of losing, etc. Schoemaker concluded that:

> variations in decision strategies . . . can, in part, be understood from such individual differences as statistical knowledge . . . the explicit use of expected values, as reported by the trained subjects themselves, strongly suggests that training itself played an important role.

This overview has given, we hope, a flavor of the diversity of studies that have been undertaken which attempt to relate individual differences in decision making to personality measures. Typical studies have looked for relationships between personality and decision making using single measures of personality and single decision situations. Various patterns of results have been obtained but little attempt has been made to see if the results generalize to other laboratory decision tasks. Next we turn to studies that have attempted to attain this much more difficult objective. The objective has been to identify distinct *cognitive styles* in decision making, one of which characterizes a specific individual over the whole range of his or her decision making.

Cognitive Style and Decision Making

Driver and Mock[24] have postulated two dimensions of information processing in decision making: the focus dimension and the amount of information utilized. There are two extremes of the focus dimension. At one pole are processors who view the data as suggesting a single course of action or solution, whereas at the other are processors who view solutions as multiple.

The amount of information used in reaching a decision is also conjectured to vary from one decision maker to another. At one extreme is the minimal data user who 'satisfices' on information use, and at the other is the maximal data user who processes all the available information that is perceived to be of use for the decision.

By combining these two dimensions of information processing in decision making, Driver and Mock derived four basic decision styles. The *decisive style* is 'one in which a person habitually uses a minimal amount of information to generate one firm option. It is a style characterized by a concern for speed, efficiency and consistency.' The *flexible style* 'also uses minimal data, but sees it having a different meaning at different times . . . It is a style associated with speed, adaptability and intuition.' In contrast, the *hierarchic style* 'uses masses of carefully analysed data to arrive at one best conclusion. It is associated with great thoroughness, precision and perfectionism.' Similarly, the *integrative style* 'uses masses of data, but will generate a multitude of possible solutions . . . It is a highly experimental, information-loving style—often very creative.' Driver and colleagues have developed two main questionnaire measures of decision style that have predicted such behavior as decision speed, use of data, information search, and information purchase on experimental tasks. However, most (if not all) of these empirical studies, including the questionnaire measures themselves, are contained in unpublished reports, and so the quality of this research is difficult to evaluate. For instance, Driver[25] has apparently shown that some persons use one style pre-dominantly, whereas others employ one style as often as another.

Other investigators[26] have published a study examining the relationship between Driver's decision styles and information processing in decision making. They used a decision situation in which their subjects were to make recommendations about whether to include certain companies in the investment portfolio of a large insurance company. Information about the companies included that on eight cue variables, including sales, net income and primary earning per share. For each set of eight cues presented, subjects were required to provide (1) a recommendation for or against further consideration of that company by their superiors; (2) an assessment of their confidence in their recommendation; and (3) an indication of the amount of additional information needed in order for the subject to increase confidence in his or her recommendation. By constructing a multiple linear regression model of each subject, these investigators were able to measure 'use of information'. However, they found no significant difference in information usage between those subjects classified as maximal or minimal data users on the basis of Driver's Integrative Style questionnaire. Also, those subjects who showed low or high Tolerance of Ambiguity on a questionnaire measure did not have significantly different levels of confidence in their recommendations or a differential desire to seek additional information in order to increase their confidence in a recommendation. These investigators concluded that cognitive style and personality variables do not appear to be useful in predicting human information processing.

The Myers–Briggs type indicator[27] has also been used to discriminate decision styles. This follows the psychology of types developed by Jung. According to Casey,[28] individuals categorized as sensors 'prefer to analyze isolated, concrete details in making a decision', whereas intuitors 'focus on relationships, or gestalt'. In this study, bank loan officers made predictions of the possible corporate failure of each of 30 firms based on the information contained in six financial ratios, such as net income/total assets for each firm. The ratios were real ones, belonging to 15 firms that had already filed for bankruptcy and to 15 randomly chosen non-bankrupt firms.

Casey describes the way in which Jung's information-processing styles were hypothesized to be related to performance on his task:

> After an explanation and discussion of Jung's theory between me and the expert panel of bankers, the panel predicted that the intuitors would perform significantly better than the sensors in the prediction task. In the panel's opinion, analysis of financial ratios to predict failure accurately required that the levels and trends of the ratios, as well as the possible trade-offs among them (e.g. liquidity for profitability), be economical (pp. 605–606).

Casey's hypothesis was confirmed in that he found a moderate correlation, in the expected direction, between overall predictive accuracy and the two information-processing types.

Another researcher[29] also utilized the Myers–Briggs instrument, this time to differentiate individuals' performance on a computer simulation of a production function. Individual decision makers acted as operations managers. One of the tasks faced by his subjects was to develop a production plan for a 5-week production period with the objective of minimizing the firm's total costs. This researcher found that sensing subjects obtained significantly lower costs than intuitive subjects. He argued that this was because his decision task was analytical and moderately well structured, whereas other tasks involving tactical and strategic decisions would be less well structured and would tend to favor good performance by intuitive types.

Wright and Phillips[30] argued that it was possible to define alternate cognitive styles of probabilistic and non-probabilistic thinking and predict performance across a variety of decision tasks. Specifically, they argued that a probabilistic thinker with no cognitive limitations would take a probabilistic rather than non-probabilistic view when confronted with uncertainty, would value information that could reduce uncertainty, would revise probabilities in light of new information, would be less prone to violate a normative axiom of decision theory, would take account of future uncertainties when making plans, and would show no bias for certain over uncertain events just because the former are certain. On the other hand, the non-probabilistic thinker would translate uncertainty into yes–no or 'don't know' terms, would put little value on fallible information, would show little revision of probabilities when fallible information was presented, would be prone to violate a normative axiom of decision theory, would make plans on the basis of best guesses, and would be biased toward opinions with certain consequences.

Wright and Phillip's predictions for performance were expressed entirely in terms of expected differences between non-probabilistic and probabilistic cognitive styles, even though they did not have an independent measure that would distinguish people adopting these styles. The purpose of their research was to discover whether these two styles were consistently adopted by people over a variety of decision and inference tasks. However, they found little evidence of cross-situational consistency of their hypothesized cognitive styles.

In summary, the overall result of attempts to delineate distinct cognitive styles of decision making that characterize individuals over a variety of decision situations or tasks has been, to date, disappointing. If we accept that there is little evidence of consistency in individual decision behavior, such that distinct cognitive decision styles *cannot* be determined with confidence, it follows that attempts to devise 'personalized' methods of presenting information or aiding decision making that can match or complement the decision styles of the decision makers may be wasted efforts. Huber[31] summarizes this point of view:

> The multitude of measuring instruments for assessing cognitive style, in combination with the variety of behaviors and performance measures that have been studied, causes there to be a limited number of precisely comparable studies . . . Thus the literature must be labeled as inconclusive, since either (1) the sparseness of comparable studies prohibits precise comparisons, or (2) the aggregation of non-comparable studies leads to the identification of apparent inconsistencies.

Huber concluded on the basis of this argument that present research is insufficiently sound to be a basis for designing 'individualized' decision support (DSS) or management information systems. He also concluded that further cognitive style research is unlikely to lead to operational guidelines for the design of such systems because

> (1) there are many individual differences related to decision-making behavior, and the task of constructing an empirically-based normative design model that accounts for all their effects is overwhelming, and (2) even if we could build such a model, it would be inapplicable to any one decision-maker because there are individual differences in the nature and extent of association among individual differences.

Huber's resolution of these problems is clear cut:

> This means that the DSS design effort should be directed towards creating a DSS that is flexible, friendly, and that provides a variety of options. If this focus is adopted, the matter of an *a priori* determination of the user's style as a basis for identifying the most appropriate design becomes largely irrelevant.

In contrast to the search for separable and consistent individual differences in decision making, other researchers have placed major emphasis on the contribution of task characteristics to decisional variance. As we saw earlier in this chapter, Pitz has demonstrated that very small changes in the task can produce evidence of poor or good judgment. We now turn to analyze in more depth the contingent nature of decision making.

CONTINGENT DECISION BEHAVIOR

Kahneman and Tversky,[32] in their 'prospect theory' of decision making, argued that the subjective valuation of the possible outcomes of a decision is contingent upon the decision maker's reference point or frame of reference:

> Framing effects in consumer behavior may be particularly pronounced in situations that have a single dimension of cost (usually money) and several dimensions of benefit. An elaborate tape deck is a distinct asset in the purchase of a new car. Its cost, however, is naturally treated as a small increment over the price of the car. The purchase is made easier by judging the value of the tape deck independently and its cost as an increment. Many buyers of homes have similar experiences. Furniture is often bought with little distress at the same time as the house. Purchases that are postponed, perhaps because the desired items were not available, often appear extravagant when contemplated separately; their cost looms large on its own. The attractiveness of a course of action may thus change if its cost or benefit is placed in a larger account.

In another study that emphasized the contingent nature of decision making, Lichtenstein and Slovic[33] had subjects evaluate gambles by a choice procedure and also by a bidding procedure. The choice procedure required subjects to indicate which of a pair of gambles they preferred, whereas the bidding procedure asked subjects to name an amount of money at which they would be indifferent between playing a specified single gamble or of having that amount of money. When they compared the results of the choice and bidding procedures they found, surprisingly, that for the same subject the results of the two procedures were not correlated. Specifically, they often found that subjects would indicate a greater preference for one gamble when a choice procedure was used and that they would bid more for another gamble when a bidding procedure was used. When asked to choose between gambles, people tended to prefer those containing a higher probability of winning, whereas higher bids were made for gambles containing larger amounts to win.

Svenson[34] has investigated the relationship between decision rules or strategies and choice between alternatives. His review demonstrated that most decision problems are solved without complete search of available information, whereas think aloud protocols, obtained from individual problem solvers, revealed that many different decision rules are used in simple choice situations. One general conclusion is that less information is obtained when the number of alternatives or of value attributes describing the alternatives is increased. In addition, there seems to be a general tendency toward increased intra-alternative search and decreased intra-attribute search when the number of alternatives in a choice set is increased. Svenson also speculated on the role of factors, other than those contained in the task, in determining decisional variance:

> Although the results seem to indicate the existence of individual search patterns in some of the investigated decision situations, little is known about the consistency of rules across situations. However thought provoking the idea, it is much too early to allege the existence of individual decision habits at present.

Payne[35] summarized most of this situation-oriented research:

> The present review strongly suggests the conclusion that decision making is a highly contingent form of information processing. The finding that decision behavior is sensitive to seemingly minor changes in task and context is one of the major results of years of decision research. It will be valuable for researchers to continue to identify task and context effects. Nevertheless, the primary focus of decision research should now be the search for some general principles from which contingent processing would follow.

Payne has identified three major theoretical frameworks for dealing with task and context effects on decision making. The first, production systems, has been exemplified earlier in the work of Gordon Pitz. Payne notes further that the decision maker's testing of conditions for applying a decision rule is considered to be automatic and unconscious. Also the production system framework is very general, and one implication is that there could be large individual differences in response to a particular task environment. A second framework, cost/benefit analysis, views choice of strategy in decision making as a conscious process. Benefits could include:

> the probability that the decision strategy will lead to a correct decision, the speed of making the decision, and its justifiability. Costs might include the information acquisition and computational effort involved in using a strategy. Decision rule selection would then involve consideration of both the costs and benefits associated with each possible strategy.

A third framework is in terms of a perceptual view. This framework is explicit in the work of Kahneman and Tversky on the framing of decisions which we outlined earlier in this section. Kahneman and Tversky argue that people are often unaware of framing effects and, once they are made aware, they are still unable to see decision problems in an objective way. The analogy is with perception research on the psychology of illusion. Because the illusory effect produced by line drawings (for example, the Muller Lyer) have been shown to be fairly consistent across subjects, the implication is that task and context effects on decision making will also be fairly consistent across subjects. Because illusory effects have been shown to be predominantly 'wired in' rather than under conscious control, it follows that the costs and benefits of a particular decision strategy should have a minor effect on strategy choice. However, more data need to be collected before the relative dominance of any one of these three frameworks is shown to be the better explanation of task influences on choice behavior.

SUMMARY

This chapter has described and evaluated research by psychologists on the quality of human judgment and decision making. Most of the research has taken place in the psychological laboratory and has, more recently, been the subject of critical comment concerning relevance to real-world contexts.

Studies of individual differences in decision making and judgment abilities have investigated possible relationships to personality characteristics of decision makers and have attempted to differentiate distinct cognitive styles of decision making that apply over varied situations. However, designers of decision support systems are sceptical about the worth of designing systems to match or complement such individual differences. Overall, decision making and judgment seem contingent on the nature of the decision-making situation.

DISCUSSION QUESTIONS

(1) Is decision making contingent on the decision faced or on the personality characteristics of the decision maker?
(2) Should Bayes' theorem be used to aid decision makers revise their opinions as they gain more information?
(3) How confident are you that the heuristics and biases that have been identified in the psychological laboratory generalize to real-world decision making?

REFERENCES

1. Much of their work is reviewed in Tversky, A. and Kahneman, D. (1974) Judgement under Uncertainty: Heuristics and Biases, in *Science*, **185**, 1124–1131.

For a collection of papers on heuristics and biases see Kahneman, D., Tversky, A. and Slovic, P. (1981) *Judgment under Uncertainty: Heuristics and Biases,* Cambridge University Press, Cambridge.
2. Lichtenstein, S., Slovic P., Fischhoff, B., Layman, M. and Coombs, B. (1978) Judged Frequency of Lethal Events, in *Journal of Experimental Psychology: Human Learning and Memory,* 4, 551–578.
3. Slovic, P. (1972) From Shakespeare to Simon: Speculations—and Some Evidence—about Man's Ability to Process Information, in Research Monograph, Vol. 12, No. 12, Oregon Research Institute, April.
4. Kahneman, D. and Tversky, A. (1977) Intuitive Prediction: Biases and Corrective Procedures, in S. Makridakis and S. C. Wheelwright (eds), *TIMS Studies: Management Science,* 20, 313–327.
5. Cohen, L. J. (1981) Can Human Irrationality be Experimentally Demonstrated? *Behavioral and Brain Sciences,* 4, 317–370: Ebbeson, E. B. and Koneci, V. J. (1980) On the External Validity of Decision Making research: What do we know about Decisions in the Real World? in T. S. Wallsten (ed.) *Cognitive Processes in Choice and Decision Making,* Erlbaum, Hillsdale, NJ.
6. Beach, L. R., Christensen-Szalanski, J. and Barnes, V. (1987) Assessing Human Judgment: Has it been Done, Can it be Done, Should it be Done? in G. Wright and P. Ayton (eds) *Judgmental Forecasting,* Wiley, New York.
7. Pitz, G. F. (1977) Decision Making and Cognition, in H. Jungermann and G. de Zeeuw (eds) *Decision Making and Change in Human Affairs,* D. Reidel, Amsterdam.
8. Kahneman, D. and Tversky, A. (1982) The Psychology of Preferences, *Scientific American,* 39, 136–142.
9. Christensen-Szalanski, J. (1986). Improving the Practical Utility of Decision Making Research, in B. Brehmer, H. Jungermann, P. Lourens and G. Sevon (eds) *New Directions in Research on Decision Making,* North-Holland, Amsterdam.
10. Phillips, L. D., (1987) On the Adequacy of Judgmental Forecasts, in G. Wright and P. Ayton (eds) *Judgmental Forecasting,* Wiley, Chichester.
11. Phillips, L. D. and Edwards, W. (1966) Conservatism in a Simple Probability Inference Task, *Journal of Experimental Psychology,* 72, 346–354.
12. Peterson, C. R., Schneider, R. J. and Miller, A. J. (1965) Sample Size and the Revision of Subjective Probability, *Journal of Experimental Psychology,* 69, 522–527.
13. Pitz, G. F., Downing, L. and Reinhold, H. (1967) Sequential Effects in the Revision of Subjective Probabilities, *Canadian Journal of Psychology,* 21, 381–393.
14. DuCharme, W. M. and Peterson, C. R. (1968) Intuitive Inference about Normally Distributed Populations, *Journal of Experimental Psychology,* 78, 269–275.
15. Winkler, R. L. and Murphy, A. M. (1973) Experiments in the Laboratory and the Real World, *Organizational Behavior and Human Performance,* 10, 252–270.
16. Youssef, Z. I. and Peterson, C. R. (1973) Intuitive Cascaded Inferences, *Organizational Behavior and Human Performance,* 10, 349–358.
17. Adorno, T. C., Frenkel-Brunswik, E., Levinson, P. J. and Sanford, R. N. (1960) *The Authoritarian Personality,* New York, Norton, p. 480.
18. Ertel, S. (1972) Erkenntnis und Dogmatismus, *Psychologische Rundschau,* 23, 241–269.
19. Wright, G. N. and Phillips, L. D. (1979) Personality and Probabilistic Thinking: An Exploratory Study, *British Journal of Psychology,* 70, 295–303.
20. Taylor, R. N. and Dunnette, M. D. (1974) Relative Contribution of Decision-maker Attributes to Decision Processes, *Organizational Behavior and Human Performance,* 12, 286–298.

21. Kogan, N. and Wallach, M. A. (1960) Certainty of Judgment and Evaluation of Risk, *Psychological Reports*, **6**, 207–213.
22. Plax, T. G. and Rosenfeld, L. B. (1976). Correlates of Risky Decision-making, *Journal of Personality Assessment*, **40**, 413–418.
23. Schoemaker, P. J. H. (1979) The Rate of Statistical Knowledge in Gambling Decisions: Moment versus Risk Dimension approaches, *Organizational Behavior and Human Performance*, **24**, 1–17.
24. Driver, M. J. and Mock, T. J. (1975) Human Information Processing, Decision Style Theory, and Accounting Systems, *The Accounting Review*, July, 490–508.
25. Driver, M. J. (1974) *Decision Style and its Measurement*, unpublished manuscript, Graduate School of Business Administration, University of Southern California, Los Angeles.
26. McGhee, W., Shields M. D. and Birnberg, J. G. (1978) The Effect of Personality on a Subject's Information Processing, *The Accounting Review*, July, 681–697.
27. Myers, I. B. (1982) *Manual: The Myers–Briggs Type Indicator*, Educational Testing Service, Princeton, NJ.
28. Casey, C. J. (1980) The Usefulness of Accounting Ratios for Subjects' Predictions of Corporate Failure: Replication and Extensions, *Journal of Accounting Research*, **18**, 603–613.
29. Davis, D. L. (1982) Are Some Cognitive Types Better Decision Makers than Others? An Empirical Investigation, *Human Systems Management*, **3**, 165–172.
30. Wright, G. N. and Phillips, L. D. (1984) Decision Making: Cognitive Style or Task-related Behavior? In H. Bonarius, G. Van Heck and N. Smid (eds) *Personality Psychology in Europe*, Swets and Zeitlinger, Lisse.
31. Huber, G. P. (1983) Cognitive Style as a Basis for MIS and DSS Designs: Much Ado about Nothing? *Management Science*, **29**, 567–579.
32. Kahneman, D. and Tversky, A. (1982) The Psychology of Preferences, *Scientific American*, **39**, 136–142.
33. Lichtenstein, S. and Slovic, P. (1971) Reversals of Preference between Bids and Choices in Gambling Decisions, *Journal of Experimental Psychology*, **89**, 46–55.
34. Svenson, O. (1983) Decision Rules and Information Processing in Decision Making, in L. Sjoberg, T. Tyszka and J. A. Wise (eds) *Human Decision Making*, Doxa, Bodafors, Sweden.
35. Payne, J. W. (1982) Contingent Decision Behavior, *Psychological Bulletin*, **92**, 382–402.

Chapter 9

The Quality of Human Judgment: Real-world Studies

INTRODUCTION

The previous chapter analyzed studies of the quality of human judgment in the artificial setting of the psychological laboratory. This chapter first extends discussion of the generalizability of such results and then evaluates studies of the quality of human judgment in real-world forecasting contexts. As we shall see, these studies have, for the most part, been conducted on the more general role of judgment in forecasting rather than on assessing subjective probabilities for the occurrence of future events. However, analysis of this work provides the necessary background for a full appreciation of the capabilities of human judgment.

Indeed, judgment is implicated in many ways with a variety of forecasting methods. This chapter briefly outlines the major methods (extrapolation and econometric modeling) and focuses on the role of judgment in forecasting with statistical models by considering anecdotal reports of the real-world use of these models. The concept of 'expertise' is discussed and we go on to analyze the implications of studies of expert judgment for the quality of judgmental 'adjustments' to results derived from statistical models in everyday situations. First, though, we extend Chapter 8's discussion of the quality of probability judgment in laboratory settings to studies of assessing probabilities for future events.

ASSESSING PROBABILITIES FOR FUTURE EVENTS

Psychological research on the quality of judgment has very fundamental implications since it is often used as the basis for generalization into forecasting practice. For example, perhaps the most commonly quoted article in the forecasting literature which expresses doubt about the capabilities of human judgment is that by Hogarth and Makridakis,[1] which was published in 1981. These authors argue that 'many of the numerous processing limitations and biases revealed in the literature apply to tasks performed in forecasting and planning' (p. 115). However, in this case, we must observe that the biases quoted had mostly, at that time, been identified in undergraduate students' answers to simple paper and pencil tasks completed in the psychological laboratory, discussed in Chapter 8. Further, most of the tasks related to judgment *per se* rather than judgment in forecasting. Hogarth and Makridakis, rather than the authors who subsequently cite their work, did recognize that these 'biases found to operate in the psychological laboratory might not generalize to more naturalistic environments.'

Similarly, when these authors conclude that forecasters have a 'mistaken confidence in judgment' they were using the early references on the calibration of subjective probability judgments.[2] As we mentioned briefly in Chapter 8, calibration is one measure of the validity of subjective probability assessment such that for a person to be perfectly calibrated, assessed probability should equal percentage correct where repetitive assessments are being used. To recapitulate, if you assign a probability of 0.8 as the likelihood of each of ten events occurring, eight of those ten events should occur if the probability forecasts are perfectly calibrated. Similarly, all events assessed as certain to occur (1.0 probabilities) should, in fact, occur. The relation between assessed probability, proportion correct and calibration is given in Figure 9.1.

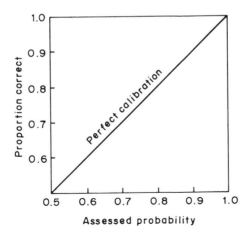

Figure 9.1 Perfect calibration

Early studies of calibration almost exclusively used general knowledge items in the form of dichotomous questions such as 'Which canal is longer, (a) the Suez Canal or (b) the Panama Canal?' Respondents were required to indicate the answer they think is correct and then to assess a probability between 0.5 to 1 to indicate their degree of belief in its correctness. General knowledge questions have been extensively used in studies of calibration because subjects' answers can be immediately and conveniently evaluated by the experimenter. This research has documented the generality of 'overconfidence' in probability assessment in that proportion correct has been found to be less than assessed probability. Figure 9.2 shows typical overconfidence found in calibration studies utilizing general knowledge questions.

It has more recently been observed that probability assessments for future events relate to different cognitive processes than those involved in putting a probability to the veracity of one's own memory. A series of studies[3,4] has shown differences in calibration and related measures for sets of questions where the answer is already known (general knowledge verification) and where the answer is not known at the time of the probability assessment (judgmental forecasting). In general, people do not use as many certainty assessments in judgmental forecasting and the forecasts tend to be much better calibrated. One instance where judgmental forecasts are routinely generated is weather forecasting. The official forecasts issued by the National

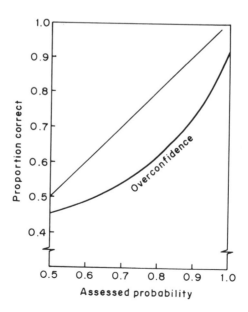

Figure 9.2 Overconfidence

Weather Service in the United States are subjective probability forecasts. Murphy and colleagues[5] have evaluated these subjective forecasts and found that, for certain categories of weather, they were more accurate than the available objective statistical techniques. In this case, the forecasters have a very large amount of information available, including the output from statistical techniques. They also receive detailed feedback and have the opportunity to gain experience of making forecasts over a wide range of meteorological conditions. Furthermore, they have considerable practice in quantifying their internal state of uncertainty. These circumstances may well be ideal for the relatively successful application of judgmental, as compared with purely quantitative, forecasting.

Additionally, good calibration has been demonstrated in several real-world forecasting situations apart from weather forecasting. These situations include horse racing,[6] prediction of the future interest rates by bankers[7] and prediction of the success of R&D projects.[8]

It appears that performance-demonstrated expertise in probability judgments is underpinned by practice and regular performance feedback. As Einhorn and Hogarth[9] have argued, most judgments are made without the benefit of accurate feedback. Einhorn traced these difficulties to two main factors. The first is a lack of search for and use of disconfirming evidence and the second is the use of unaided memory for coding, sorting and retrieving outcome information.

Other investigators have argued[10] that the presence of actual or potential users of judgmental forecasts provides the forecasters with a strong motivation for conducting the forecasting process in an efficient and more effective manner. Moreover, feedback from users of forecasts frequently contains information regarding possible improvements. The use of judgment in real-world forecasting thus contrasts strongly with the study of judgment in the psychological laboratory.

On this experimental versus real-world theme it is interesting to trace a parallel pattern in psychological research on human judgment within the behavioral decision-making literature. As we have shown, early research on calibration of subjective probabilities showed evidence of overconfidence bias. This result was generalized to judgmental probability forecasting.[11] Later research questioned the validity of this generalization on the grounds that the tasks were contrived laboratory settings. As we saw in Chapter 8, a similar issue can be identified in Beach et al.'s critique[12] of other judgmental biases and Winkler and Murphy's critique[13] of laboratory-based studies of human revision of opinion. In the behavioral literature this issue is known as 'ecological validity'[14] and refers to the extent to which conclusions from the laboratory really do generalize into their real-world settings. Many examples of systematically different laboratory and real-world behavioral patterns have been identified.

Such recognition of the need to preserve ecological validity in laboratory studies of judgment links closely to the study of people operating within their own expertise, a topic we shall turn to next when we will evaluate the quality of human judgment in real-world forecasting contexts. Here statistical models are commonly utilized. First, we turn to statistical extrapolation and then to econometrics. We provide a brief outline of each technique and describe anecdotal reports of judgmental interventions.

STATISTICAL EXTRAPOLATION

This type of forecasting method is based solely on the variable to be forecast (e.g. historic sales of a product) and time (e.g. on a month-by-month basis). Since only two variables are involved and since elapsed time is not the cause (or at least not the sole cause) of sales, plotted values of the two variables can only indicate the relationship between them. Typical methods of measuring the form of this relationship and using this measurement to predict the future relationship include simple linear regression, moving averages and exponential smoothing. Extrapolation methods cannot predict 'turning points' where, for example, unique changes in the world have a *causal* impact on the variable to be forecast.

There are many anecdotal reports that reveal the ubiquity of judgment in extrapolation. For example, Lawson[15] discusses traffic usage forecasting in telecommunications at the Bell Telephone Company of Pennsylvania. The quantities of switching equipment that Bell own today should cover their needs for the subsequent two years. However, Bell must buy to serve traffic loads that they expect for up to 5 years hence. On the surface, Bell's methodology for forecasting focuses on statistical extrapolation techniques, but, as Lawson notes:

> If the projections are acceptable, based on engineering equipment, they are accepted. If not the forecaster decides which historical data points contributed to the bad trend, deletes those historical values, and recycles the process until an acceptable result is obtained. When all else fails the forecaster abandons the mathematics and substitutes an acceptable forecast based on judgment.

Further:

> The eyeball method involves the forecaster in the analysis of data at all steps in the process. It avoids the trap of mathematical projection systems—the tendency to believe the fitted trend without analysis of whether or not the historical data points are representative. In practice it has produced better forecasts than more statistically based systems . . . The eyeball method is better because it involves the forecaster in analysis of the historical data as the projection is made.

In the context of sales forecasts, one commentator[16] argues that:

> . . . while the computer makes it easy to use sophisticated tools, executive judgment will always be needed to assess how the future might differ from the past, and to adjust the forecast accordingly . . .

Also in sales forecasting, another commentator[17] points out that:

> Quantitatively advanced techniques such as regression modeling, Box–Jenkins, exponential smoothing . . . cannot anticipate one-time events such as surprise competitive development nor are they particularly effective for long term planning without significant adjustment by management judgment.

ECONOMETRICS

This approach is based on a representation of causal relationships between variables. Often, multiple linear regression techniques can be used to model the relationships between changes in precursor variables and their effect on a target variable, where the former are presumed to have a causal impact.

Large econometric models, such as models of a nation's economy, are, in effect, hypotheses about the way in which the economy works. The Wharton model, for example, uses 400 equations which model relationships between variables. Users of such models can simulate the effects of different policy proposals, e.g. the effect of a two-point cut in the standard rate of income tax on consumer spending. Since the models are hypotheses it follows that alternative models may disagree about fundamentals. For example, the Wharton model links consumer spending closely to changes in disposable personal income whereas the Federal reserve model does not. Sometimes the forecast of the econometric model may make little sense to the modeler:

> . . . the Wharton model, like just about all others is endlessly subject to adjustment. The notion that models are challenging human judgment has some validity to it some of the time. But a lot of the time the process is reversed— i.e. the models' findings are themselves being challenged and overridden.[18]

Other investigators report on economic forecasting by the Maine State Planning Office. Here a 10-year forecast was built around a model of the state's economy. They noted that:

> . . . the model forecast substantial increases in construction employment. Unfortunately, recent growth (in the period 1975–1980) was almost entirely attributable to massive rebuilding of the paper industry's physical plants and a large spurt in Federally subsidized water, sewer and housing construction.

> Neither seemed likely to continue . . . the model's output had to be 'managed' through the application of operator judgment and experience.[19]

A further point:

> For various understandable reasons, the data on which forecasting is based are often imperfect . . . This is either because the basic data are not readily available or because those submitting them do not go to the effort or expense of ensuring their accuracy . . . the use of forecasts of future levels of national output etc requires exercise of considerable judgment.[20]

Most judgmental adjustments to econometric models are made for two reasons:[21]

(1) The model has not been performing adequately recently, and/or
(2) Some external factor, not incorporated into the model, is expected to influence future events.

In summary, the results of statistical extrapolation and econometric modeling are routinely subject to *post hoc* 'adjustment' by managers and modelers alike. These adjustments have face validity to the forecasters themselves.

Indeed, surveys of corporate forecasting practices show that most important forecasts involve judgment. One survey of 52 manufacturing firms found that 50 of them used judgmental methods in one form or another.[22] Another survey[23] of 500 of the world's largest corporations found that the overwhelming majority of corporate planners identified severe limitations in using pure statistical techniques. In the context of sales forecasting, a further survey[24] of 100 companies, drawn from the *Fortune 500* firms, found that 89% used judgmental forecasting alone or combined with other methods and 28% used judgment only. These researchers defined judgment as 'opinion' rather than judgmental *adjustment* to statistical models. However, in all these surveys there is a lack of precision in specifying the actual nature of the judgmental assessments and the manner in which they interact with quantitative methods.

In the next section we analyze the research studies that have been undertaken on the quality of judgment in forecasting. We turn first to statistical extrapolation and then to econometrics.

JUDGMENT IN FORECASTING

As we have seen, practitioner reports have tended to endorse the use of judgment. Two groups of researchers[25] have demonstrated that when an unusual event, like a promotion, occurs, time-series sales forecasts combined with judgment will substantially increase forecasting accuracy for the atypical period. These results are congruent with practitioners' advice. A later study,[26] however, utilized extrapolation methods modified with two variables which 'retailers often express' as having a major impact on sales, i.e. number of shopping days between Thanksgiving and Christmas and the number of weekends in a month, and found that these adjustments had negligible impact on the accuracy of the sales forecasts. Clearly, untested 'conventional wisdom' can be, at best, irrelevant and, at worst, misleading.

Angus-Leppon and Fatseas[27] had a final-year undergraduate accounting students forecast short-term interest rates using various judgmental methods and statistical forecasting techniques. Their experiment was a well-designed one in several aspects. First, they had the students make forecasts using judgmental and statistical methods where no information on the nature of the time series was provided (which we shall term the *abstract* condition) and second, where the subjects had increased information and knowledge about the nature of the forecast series (which we will term the *informed* condition). The informed condition is, of course, more representative of real-world forecasting practice. The judgmental methods included eyeball extrapolation of data plotted graphically and extrapolation of data presented in tabular format. Presentation of data in a *tabular* format tends to mask seasonal factors and so, as might intuitively be expected, judgmental forecasting using *graphical* methods tended to outperform judgmental forecasts based on tabular data. The judgmental forecasts of these inexperienced forecasters over a 6-month period were quite accurate, with a mean absolute percentage error of about 8%. *Knowledge* of the series being forecast *did not* improve forecast accuracy for these *inexperienced* forecasters, but their judgmental forecasts were little worse than the best statistical model employed, the fitting of a power curve.

In contrast to the above study, several recent papers have focused on what we have called abstract forecasting tasks. For example, in a comprehensive study, Lawrence *et al.*[28] compared judgmental forecasts to statistical models using abstract data. Using 111 different real-life time series, they found that judgmental extrapolative forecasting was as accurate as a range of statistical techniques and more accurate in some instances. In another experiment, Lawrence[29] used eight time series concerning 'sale of manufactured items' which exhibited 'seasonal variations and significant randomness'. Eyeball forecasts were overall the most accurate compared to single and double exponential smoothing and the Box–Jenkins technique. In this case, the condition was also abstract, since the forecasters had no prior knowledge of the time series, and Lawrence recognized the limitation of this and noted that: 'Such knowledge should enable the forecaster to do a better job and improve the judgmental forecast.'

In a later study, attempting to provide a common basis for both types of forecast, Lawrence et al.[30] used another abstract forecasting task, involving 68 monthly time series. The focus of this later study was the combination of judgmental and statistical forecasts by simple averaging as the 'simplest way in which an executive could combine two forecasts'. Lawrence et al. concluded that, overall, a combination of statistical and judgmental forecasts resulted in an improvement of accuracy over the constituent forecasts and that of combining statistical forecasts.

Carbone et al.[31] reported an experiment comparing extrapolation with judgmental forecasting. These researchers used 25 time series. They had small teams of MBA students prepare forecasts using the statistical methods. In addition, the teams had later to prepare judgmental forecasts on the basis of the 'information available to them'. This consisted of their previously made statistical forecasts using the above methods. Carbone et al. concluded that judgmental adjustment by the students did not improve accuracy, but noted that the student sample did not have expert knowledge in the series they examined.

In contrast to the majority of research studies cited above, one major review has focused on what we have termed *informed* forecasting tasks. In this work Armstrong[32] investigated annual share earnings forecasts. He noted that the most popular methods to forecast earnings are judgment (by either management or outside analysts) and extrapolation. Previous research on the accuracy of judgment and extrapolation in this context was mixed. Of eight studies reviewed by Armstrong, two concluded judgment was superior, once concluded extrapolation was superior and five were inconclusive. Armstrong's analysis focused on two types of judgmental forecast of share earnings: those from management and those from analysts. He concluded that experts, including management and analysts, have valuable information on a firm's current status and that this information helps them make better forecasts than those obtained from extrapolations.

Thus the value of judgmental forecasts is endorsed by both researcher and practitioner. The evidence is equivocal where novices, uninformed subjects, make judgmental forecasts (i.e. the abstract experiments) but where expert, informed and involved subjects use their judgment, the weight of published evidence seems to support its comparative value.

Perhaps the relative success of judgment compared with extrapolation models in these studies is due to the simplicity of the optimal forecasts. For example, one researcher re-analyzed a large number of time-series forecasts and found that a simple model which forecast the next period as the value of the last period (seasonally adjusted) was superior to more sophisticated statistical extrapolation methods. In addition, it is clear that a considerable amount of averaging in the results of these large-scale studies will tend to favor simpler methods because they are more robust and generally applicable, while sophisticated methods are more specific in their applications. However, if the reason that judgment is relatively successful in univariate time-series extrapolation is due to the simplicity of optimal forecasts, then this result should not generalize when judgment is compared with multivariate econometric systems.

As we have seen, the output of econometric models are routinely subject to judgmental adjustment. Given this fact, Armstrong[33] asked why most textbooks imply that econometric methods provide more accurate *short-range* forecasts than others. He mailed a questionnaire to 55 experts in econometrics based at leading American business schools; 21 responses were returned, of which 95% agreed with a statement that predictions from econometric models were more accurate than forecasts 'obtained from competitive models for short-term forecasting'. Next, Armstrong examined the existing empirical studies that compared econometric predictions to extrapolative predictions and found that only one showed that the econometric model provided superior predictions. In the four that compared econometric versus purely judgmental forecasts, he found no evidence of statistically significant superiority, and when he compared simple versus complex econometric methods, he found that seven out of 11 comparisons favored less complexity and four favoured more complexity. Thus the published evidence suggests that purely judgmental forecasts appear to perform as well as the econometric models. This is consistent with the survey of univariate techniques above.

SUMMARY

This chapter has documented the validity of judgment in contexts akin to real-world forecasting practice. Overall, holistic or expert judgment performs well in comparison to statistical forecasting models. This result confirms the (as then unverified) conclusion of Chapter 8: human judgment can be of high quality, particularly in real-world contexts. However, the practice of informal adjustments to the output of quantitative models, although commonplace, may be difficult to defend and communicate to colleagues or senior management. Our view is that, within the practice of decision analysis, the structured decomposition and recomposition of judgment and subsequent sensitivity analyses allow a more formal interaction between judgment and analytical model[34]. As we have seen in earlier chapters, sensitivity analysis on judgmental inputs to decision analysis allow identification of *critical* judgmental inputs for subsequent focused evaluation.

DISCUSSION QUESTIONS

(1) How is judgment involved in forecasting?

(2) To what extent do studies of judgment in forecasting practice provide evidence of good judgmental performance?

(3) Should the role of judgment in forecasting be minimized such that the outputs of statistical forecasting models are utilized whenever possible?

REFERENCES

1. Hogarth, R. M. and Makridakis, S. (1981) Forecasting and Planning: An Evaluation, *Management Science*, **227**, 115–138.
2. Lichtenstein, S., Fischhoff, B. and Phillips, L. D. (1977) Calibration of probabilities: The State of the Art, in H. Jungermann and G. de Zeeuw (eds) *Decision Making and Change in Human Affairs*, D. Reidel, Amsterdam: Lichtenstein, S., Fischhoff, B. and Phillips, L. D. (1982) Calibration of probabilities: The State of the Art to 1980, in D. Kahneman, P. Slovic and A. Tversky (eds) *Judgment under Uncertainty: Heuristics and Biases*, Cambridge Univeristy Press, Cambridge.
3. Wright, G. (1982) Changes in the Realism and Distribution of Probability Assessment as a Function of Question Type, *Acta Psychologica*, **52**, 165–174: Wright, G. and Wisudha, A. (1982) Distribution of Probability Assessments for Almanac and Future Event questions, *Scandinavian Journal of Psychology*, **23**, 219–224.
4. Wright, G. and Ayton, P. (1986) Subjective Confidence in Forecasts: a Reply to Fischhoff and McGregor, *Journal of Forecasting*, **5**, 117–123: Wright, G. and Ayton, P. (1988) Immediate and short-term Judgemental Forecasting: Personologism Situationism, or Interactionism? *Personality and Individual Differences*, **9**, 109–120.
5. Murphy, A. H. and Brown, B. G. (1985) A Comparative Evaluation of Objective and Subjective Weather Forecasts in the United States, in G. Wright (ed.) *Behavioural Decision Making*, Plenum, New York.
6. Hoerl, A. and Falein, H. K. (1974) Reliability of Subjective Evaluation in a High Incentive Situation, *Journal of the Royal Statistical Society*, **137**, 227–230.
7. Kabus, I. (1976) You Can Bank on Uncertainty, *Harvard Business Review*, May–June, 95–105.
8. Balthasar, H. U., Boschi, R. A. A. and Menke, M. M. (1978) Calling the Shots in R and D, *Harvard Business Review*, May–June, 151–160.
9. Einhorn, H. J. and Hogarth, R. (1978) Overconfidence in Judgment: Persistence of the Illusion of Validity, *Psychological Review*, **85**, 395–476.
10. Murphy, A. H. and Brown, B. G. (1985) A Comparative Evaluation of Objective and Subjective Weather Forecasts in the United States, in Wright, G. (ed.) *Behavioural Decision Making*, Plenum, New York.
11. Fischhoff, B. and McGregor, D. (1982) Subjective Confidence in Forecasts, *Journal of Forecasting*, **1**(2), 155–172.
12. Beach, L. R., Christensen-Szalanski, J. and Barnes, V. (1987) Assessing Human Judgment: Has It Been Done, Can It Be Done, Should It Be Done? in G. Wright and P. Ayton, (Eds) *Judgemental Forecasting*, Wiley, Chichester.
13. Winkler, R. L. and Murphy, A. M. (1973) Experiments in the Laboratory and the Real World, *Organisational Behavior and Human Performance*, **10**, 252–270.

14. Willems, E. P. (1965) An Ecological Orientation in Psychology, *Merrill–Palmer Quarterly of Behavior and Development*, **11**, 317–343.
15. Lawson, R. W. (1981) Traffic Usage Forecasting: Is it an Art or a Science? *Telephony*, February, 19–24.
16. Soergel, R. F. (1983) Probing the Past for the Future, *Sales and Marketing Management*, **130**, 39–43.
17. Jenks, J. M. (1983) Non-computer Forecasts to Use Right Now, *Business Marketing*, **68**, 82–84.
18. Malley, D. D. (1975) Lawrence Klein and his Forecasting Machine, *Fortune*, March, 152–157.
19. Irland, L. C., Colgan, C. S. and Lawton, C. T. (1984) Forecasting a State's Economy: Maine's Experience, *The Northeast Journal of Business*, **11**, 7–19.
20. Glendinning, R. (1975) Economic Forecasting, *Management Accounting*, **11**, 409–411.
21. McNees, S. K. and Perna, N. S. (1981) Forecasting Macroeconomic Variables: An Electronic Approach, *New England Economic Journal*, May–June, 15–30.
22. Rothe, J. T. (1978) Effectiveness of Sales Forecasting Methods, *Industrial Marketing Management*, April, 114–118.
23. Klein, H. E. and Linneman, R. E. (1984) Environmental Assessment: An International Study of Corporate Practice, *Journal of Business Strategy*, **5**, 66–84.
24. Cerullo, M. J. and Avila, A. (1975) Sales Forecasting Practices: A Survey, *Managerial Planning*, **24**, 33–39.
25. Pankoff, L. D. and Roberts, H. V. (1968) Bayesian Synthesis of Clinical and Statistical Prediction, *Psychological Bulletin*, **70**, 762–773: Reinmuth, J. E. and Guerts, M. D. (1972) A Bayesian Approach to Forecasting Atypical Situations, *Journal of Marketing Research*, August.
26. Guerts, M. D. and Kelly, J. P. (1986) Forecasting Retail Sales Using Alternative Models, *International Journal of Forecasting*, **2**, 261–272.
27. Angus-Leppan, P. and Fatseas, V. (1986) The Forecasting Accuracy of Trainee Accountants Using Judgemental and Statistical Techniques, *Accounting and Business Research*, Summer, 179–188.
28. Lawrence, M. J., Edmonson, R. H. and O'Connor, M. J. (1985) An Examination of the Accuracy of Judgemental Extrapolation of Time Series, *International Journal of Forecasting*, May, 14–25.
29. Lawrence, M. J. (1983) The Exploration of Some Practical Issues in the Use of Quantitative Forecasting, *Journal of Forecasting*, **2**, 169–179.
30. Lawrence, M. J., Edmonson, R. H. and O'Connor, M. J. (1986) The Accuracy of Combining Judgemental and Statistical Forecasts, *Management Science*, **32**, 1521–1532.
31. Carbone, R., Anderson, A., Corriveau, Y. and Corson, P. P. (1983) Comparing for Different Time Series Methods the Value of Technical Expertise, Individualized Analysis, and Judgemental Adjustment, *Management Science*, **79**, 559–566.
32. Armstrong, J. S. (1981) What to Ask about Managements' Forecasts, *Directors and Boards*, **6**, 20–26.
33. Armstrong, J. S. (1978) Forecasting with Econometric Models: Folklore versus Fact, *Journal of Business*, **51**, 549–564.
34. See also Bunn, D. and Wright, G. (1991) Interaction of Judgmental and Statistical Forecasting: Issues and Analysis, *Management Science*.

Chapter 10

Probability Assessment

INTRODUCTION

We have seen in earlier chapters that subjective probabilities provide a concise and unambiguous measure of uncertainty and they are therefore an important element of many decision models. A number of techniques have been developed to assist the decision maker with the task of making probability judgments and in this chapter we will examine some of the more widely used methods.

It has been found that these techniques are employed most effectively when they are administered by an analyst who uses them as part of an interview with the decision maker (see Spetzler and Stäel von Holstein[1]). Of course, the assessment process can be time consuming and therefore the importance of sensitivity analysis cannot be overstated. It may reveal, for example, that a very rough assessment of the probabilities in a problem is adequate because the choice of a given course of action is valid for a wide range of probability values. Towards the end of the chapter we will examine the calibration of subjective probability estimates in depth and discuss additional ways in which the process of probability assessment can be improved.

PREPARING FOR PROBABILITY ASSESSMENT

Because of the danger of bias in probability assessment it is a good idea for the analyst to prepare the ground carefully before eliciting the probabilities themselves. Spetzler and Stäel von Holstein[1] recommend that the interview carried out by the analyst should involve three phases before the probabilities are quantified: motivating, structuring and conditioning.

Motivating

This phase is designed to introduce the decision maker to the task of assessing probabilities and to explain the importance and purpose of the task. Sensitivity analysis should be used by the analyst to identify those probabilities which need to be assessed with precision. At this stage the possibility that assessments may be subject to deliberate biases should be explored (e.g. a manager may overestimate the probability of high costs in the hope that when costs turn out to be low he will be seen in a good light). Deliberate bias is, of course, an undesirable characteristic of an input to an analysis whose intended outcome is improved decision making.

Structuring

In the structuring phase the quantity to be assessed should be clearly defined and structured. For example, asking the decision maker vague questions about 'the value of shares in the USA in 1992' is unlikely to lead to reliable responses. 'The value of the Dow Jones index at the end of trading on Wednesday 1 July 1992' is obviously a less ambiguous quantity. It is also important at this stage to agree on a scale of measurement which the decision maker feels comfortable with: if he thinks of sales in terms of 'numbers of boxes sold per week' it would be inappropriate to force him or her to assess a probability distribution for 'the number of tons sold per week'.

When the decision maker thinks that the quantity to be assessed depends on other factors it may be simpler to restructure the assessment task, possibly by making use of a probability tree (see Chapter 3 and the last section of this chapter). For example, it may be that the development time for a new product will depend upon whether two companies can agree to collaborate on the basic research. In this case the decision maker will probably find it easier to give two separate assessments: one based on the assumption that the collaboration takes place and the other on the assumption that it does not.

Conditioning

The objective of this phase is to identify and thereby avoid the biases which might otherwise distort the decision maker's probability assessments. It involves an exploration of how the decision maker approaches the task of judging probabilities. For example, are last year's sales figures being used as a basis for this year's estimates? If they are, there may be an anchoring effect. To what extent are the assessments based too heavily on the information which is most readily available, such as yesterday's news, without taking a longer-term view? More generally, the heuristics and biases identified in Chapter 8 should be borne in mind by the decision analyst as he works through the assessment process with the decision maker.

ASSESSMENT METHODS

A number of different methods have been developed for assessing probabilities. Some of these require a direct response from the decision maker in the form of a probability or quantity, while others allow the probability to be inferred by observing the decision maker's choice between bets.

Assessment Methods for Individual Probabilities

Direct Assessments

The simplest way to elicit a probability from a decision maker is to pose a direct question such as 'What is the probability that the product will achieve a break-even sales level next month?' Unfortunately, many people would feel uncomfortable with this sort of approach, and they might be tempted to give a response without sufficient thought. Asking the individual to mark a point on a scale which runs from 0 to 1 might be preferred because at least the scale enables the probability to be envisaged. Other people prefer to give their assessments in terms of odds which can then be easily converted to probabilities, as we showed in Chapter 3. For example, odds of 25 to 1 against the occurrence of an event are equivalent to a probability of 1/26 or 0.038.

The Probability Wheel

A probability wheel is a device like that shown in Figure 10.1, and it consists of a disk with two different colored sectors, whose size can be adjusted, and a fixed pointer. To see how the device might be used, let us suppose that a manager needs to assess the probability that a rival will launch a

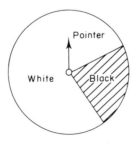

Figure 10.1 A probability wheel

competing product within the next week. We could adjust the wheel so that the white sector takes up 80% of its area and ask her to choose between the following two hypothetical gambles:

Bet One: If the rival launches the product within the next week you will win $100 000. If the rival does not launch the product you will win nothing.
Bet Two: If, after spinning the wheel once, the pointer is in the white sector you will win $100 000. If it is pointing toward the black sector you will win nothing.

If the manager says that she would choose *Bet Two* then this implies that she thinks that the probability of the rival launching the product in the next week is less than 80%. The size of the white sector could then be reduced and the question posed again. Eventually, the manager should reach a point where she is indifferent between the two bets. If this is achieved when the white sector takes up 30% of the wheel's area, this clearly implies that she estimates that the required probability is 0.3.

Note that the use of the probability wheel allowed an assessment to be made without directly asking the manager to state the probability. It is therefore an example of an indirect assessment method. The wheel also has the advantage that it enables the decision maker to visualize the chance of an event occurring. However, because it is difficult to differentiate between the sizes of small sectors, the probability wheel is not recommended for the assessment of events which have either a very low or very high probability of occurrence (we will deal with this issue later). The analyst should also ensure that the rewards of the two bets are regarded as being equivalent by the decision maker. For example, if in Bet One above, $100 000 will be paid if the rival launches within the next *year* then this would imply that the decision maker would have to wait a year before any winnings could be paid. She would probably regard this as being less attractive than a bet on the probability wheel where any winnings would be paid instantly. It is also a good idea to use a large monetary prize in the bets so that the preference between them is not influenced by other attributes which may be imposed by the assessor. The large payoff gives the monetary attribute a big weight compared to the others.

A number of devices similar to the probability wheel have also been used in probability assessment. For example, the decision maker may be asked to imagine an urn filled with 1000 colored balls (400 red and 600 blue). He or she would then be asked to choose between betting on the event in question occurring or betting on a red ball being drawn from the urn (both bets would offer the same rewards). The relative proportion of red and blue balls would then be varied until the decision maker was indifferent between the two bets, at which point the required probability could be inferred.

Assessment Methods for Probability Distributions

The Probability Method

There is evidence[2] that, when assessing probability distributions, individuals tend to be overconfident, so that they quote too narrow a range within which they think the uncertain quantity will lie. Some assessment methods fail to counteract this tendency. For example, if a decision maker is asked initially for the median value of the distribution (this is the value which has a 50% chance of being exceeded) then this can act as an anchor. As we saw in Chapter 8, it is likely that he will make insufficient adjustments from this anchor when assessing other values in the distribution. For example, the value which has only a 10% chance of being exceeded might be estimated to be closer to the median than it should be, and the result will be a distribution which is too 'tight'. Because of this, the following procedure,[3] which we will refer to as the probability method, is recommended:

Step 1: Establish the range of values within which the decision maker thinks that the uncertain quantity will lie.
Step 2: Ask the decision maker to imagine scenarios that could lead to the true value lying *outside* the range.
Step 3: Revise the range in the light of the responses in Step 2.
Step 4: Divide the range into six or seven roughly equal intervals.
Step 5: Ask the decision maker for the cumulative probability at each interval. This can either be a cumulative 'less than' distribution (e.g. what is the probability that the uncertain quantity will fall below each of these values?) or a cumulative 'greater than' (e.g. what is the probability that the uncertain quantity will exceed each of these values?), depending on which approach is easiest for the decision maker.
Step 6: Fit a curve, by hand, through the assessed points.
Step 7: Carry out checks as follows.
(i) Split the possible range into *three* equally likely intervals and find out if the decision maker would be equally happy to place a bet on the uncertain quantity falling in each interval. If he is not, then make appropriate revisions to the distribution.
(ii) Check the modality of the elicited distribution (a mode is a value where the probability distribution has a peak). For example, if the elicited probability distribution has a single mode (this can usually be recognized by examining the cumulative curve and seeing if it has a single inflection), ask the decision maker if he does have a single best guess as to the value the uncertain quantity will assume. Again revise the distribution, if necessary.

Graph Drawing

Graphs can be used in a number of ways to elicit probability distributions. In one approach the analyst produces a set of graphs, each representing a different probability density function (pdf), and then asks the decision maker to select the graph which most closely represents his or her judgment. In other approaches the decision maker might be asked to draw a graph to represent either a probability density function or a cumulative distribution function (cdf). Some computer packages (e.g. PREDICT!—see Chapter 6) allow the user to draw the shape of a probability distribution on the screen.

The *method of relative heights* is one well-known graphical technique that is designed to elicit a probability density function. First, the decision maker is asked to identify the most likely value of the variable under consideration and a vertical line is drawn on a graph to represent this likelihood. Shorter lines are then drawn for other possible values to show how their likelihoods compare with that of the most likely value.

To illustrate the method, let us suppose that a fire department has been asked to specify a probability distribution for the number of emergency calls it will receive on a public holiday. The chief administrator of the department considers that two is the most likely number of calls. To show this, the analyst draws on a graph a line which is 10 units long (see Figure 10.2). Further questioning reveals that the administrator thinks that three requests is about 80% as likely as two, so this is represented by a line eight units long. The other lines are derived in a similar way, so that the likelihood of seven requests, for example, is considered to be only 10% as likely as two and it is thought to be extremely unlikely that more than seven requests

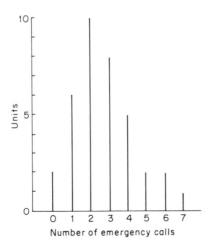

Figure 10.2 The method of relative heights

will be received. To convert the line lengths to probabilities they need to be normalized so that they sum to one. This can be achieved by dividing the length of each line by the sum of the line lengths, which is 36, as shown below (note that the probabilities do not sum to exactly one because of rounding).

Number of requests	Length of line	Probability
0	2	2/36 = 0.06
1	6	6/36 = 0.17
2	10	10/36 = 0.28
3	8	8/36 = 0.22
4	5	5/36 = 0.14
5	2	2/36 = 0.06
6	2	2/36 = 0.06
7	1	1/36 = 0.03
	36	1.00

The method of relative heights can also be used to assess probability density functions for continuous distributions. In this case the analyst will normally elicit the relative likelihood of a few values and then fit a smooth pdf curve across the tops of the lines.

A COMPARISON OF THE ASSESSMENT METHODS

Which method of probability assessment is the best? A number of experiments have been carried out to compare the methods, but these have not identified one single best method (see, for example, Seaver et al.[4] or Wallsten and Budescu[5]). Indeed, the main implication of these studies is that a variety of different methods should be used during the elicitation process. Nevertheless, certain types of approach will obviously be more acceptable than others to particular individuals. For example, some people may be happy to give direct assessments while others will prefer the indirect approach.

Bunn and Thomas[6] argued that the use of devices such as probability wheels might be most appropriate for 'people who are generally intolerant of ambiguity, those who prefer not to contemplate, or even deny, the uncertainties of their judgment, or who do not accept that psychological feelings can be quantified'. Indeed, Spetzler and Stäel von Holstein[1] found that many people have difficulty in making direct judgments, and even those who feel capable of assessing probabilities in this way were often subsequently found to lack confidence in their responses. Most subjects, they claim, prefer the probability wheel. For this reason, they recommend that interviews should generally start with assessments based on this device, while other methods should be used at a later stage as consistency checks (by consistency checks we mean testing to see if an assessment obtained by different methods or approaches is firm). Our view is that when a probability distribution has to be assessed the *probability method* is usually the best approach to adopt because, as we argued earlier, it tends to overcome the propensity of decision makers to estimate distributions which have too narrow a range.

CONSISTENCY AND COHERENCE CHECKS

Consistency checks are, of course, a crucial element of probability assessment. The use of different assessment methods will often reveal inconsistencies that can then be fed back to the decision maker. These inconsistencies should act as a stimulant to more intense thought which, hopefully, will result in greater insight and improved judgment. *Indeed, the axioms of probability theory give no guidance as to which is the best method for the elicitation of subjective probability.* Empirical research in the psychological laboratory has shown that sometimes the indirect methods are inconsistent with direct methods and sometimes they are not. Some investigators have demonstrated consistency between probability estimates inferred from wagers and direct estimates.[7] Others have shown that statistically naive subjects were inconsistent between direct and indirect assessment methods, whereas statisticians were not.[8] Generally, direct odds estimates, perhaps because they have no upper or lower limit, tend to be more extreme than direct probability estimates. If probability estimates derived by different methods for the same event are inconsistent, which method should be taken as the true index of degree of belief?

One way to answer this question is to use a single method of assessing subjective probability that is most consistent with itself. In other words, there should be high agreement between the subjective probabilities, assessed at different times by a single assessor for the same event, given that the assessor's knowledge of the event is unchanged. Unfortunately, there has been relatively little research on this important problem. One review evaluated the results of several studies using direct estimation methods. Test–retest correlations were all above 0.88 with the exception of one study using students assessing odds—here the reliability was 0.66. It was concluded that most of the subjects in all experiments were very consistent when using a single assessment method.

The implications of this research for decision analysis are not clear cut. The decision analyst should be aware that different assessment techniques are likely to lead to different probability forecasts when these are converted to a common metric.

One useful *coherence* check is to elicit from the decision maker not only the probability that an event will occur but also the probability that it will not occur. The two probabilities should, of course, sum to one. Another variant of this technique is to decompose the probability of the event not occurring into the occurrence of other possible events. If the events are seen by the probability assessor as mutually exclusive then the addition rule (Chapter 3) can be applied to evaluate the coherence of the assessments. Such checks are practically useful and are reinforced by the results of laboratory-based empirical studies of subjective probability assessment, where subjective probabilities attached to sets of mutually exclusive and exhaustive events have often been shown to sum to less than or more than one. For example, in a probability revision task, involving the updating of opinion in the light of new information, one set of researchers found four out of five subjects assessed probabilities that were greater than unity.[9] These four subjects increased their probability estimates for likely hypotheses but failed to decrease probabilities attached to unlikely hypotheses. Another probability revision study found that 49 out of 62 subjects gave probability estimates for complementary events that summed to more than unity.[10] Conversely, another investigator asked subjects to estimate sampling distributions from binomial populations on the basis of small samples, and found that in most cases subjective probabilities summed to less than unity.[11]

In a study addressed directly to the descriptive relevance of the additivity axiom, Wright and Whalley[12] found that most untrained probability assessors followed the additivity axiom in simple two-outcome assessments, involving the probabilities of an event happening and not happening. However, as the number of mutually exclusive and exhaustive events in a set was increased, more subjects, and to a greater extent, became supra-additive in that their assessed probabilities tended to add to more than one. With the number of mutually exclusive and exhaustive events in a set held constant, more subjects were supra-additive, and supra-additive to a greater degree, in the assessment of probabilities for an event set containing individuating information. In this study the individuating background information was associated with the possible success of a racehorse in a race that was about to start. It consisted simply of a record of that horse's previous performances. It seems intuitively reasonable that most probabilistic predictions are based, in the main, on one's knowledge and not to any large extent on abstract notions such as additivity. Individuating information about the likelihood of an event's occurrence may psychologically disassociate an event from its event set. As we saw earlier in Chapter 8, a similar phenomenon has been noted by Kahneman and Tversky[13] and the term 'representativeness' was coined to refer to the dominance of individuating information in intuitive prediction.

Clearly, judgmental forecasts should be monitored for additivity and incoherence should be resolved. However, a simple normalization may not be a quick and easy solution to incoherence. Lindley et al.[14] outlined a major problem:

> Suppose that I assess the probabilities of a set of mutually exclusive and exhaustive events to be
>
> $$0.001, 0.250, 0.200, 0.100, 0.279 \ldots$$
>
> It is then pointed out to me that these probabilities sum to 0.830 and hence that the assessment is incoherent. If we use the method . . . with the probability metric, we have to adjust the probabilities by adding 0.034 to each $(=(1/5)(1-0.830))$ to give
>
> $$0.035, 0.284, 0.234, 0.134, 0.313$$
>
> The problem is with the first event, which I originally regarded as very unlikely, has had its probability increased by a factor of 35! Though still small it is no longer smaller than the others by two orders of magnitude.

Obviously, other methods of allocating probability shortfalls can be devised, but our view is that the best solution to such problems is for the decision analyst to show the decision maker his or her incoherence and so allow *iterative* resolution of departures from this (and other) axioms of probability theory. Such iteration can involve the analyst plotting the responses on a graph (e.g. as a cumulative distribution function) and establishing whether the decision maker is happy that this is an accurate reflection of his or her judgments. Finally, the decision maker can be offered a series of pairs of bets. Each pair can be formulated so that the respondent would be indifferent between them if he or she is behaving consistently with assessments which were made earlier.

ASSESSMENT OF THE VALIDITY OF PROBABILITY FORECASTS

As we saw in Chapter 9, a major measure of the validity of subjective probability forecasts is calibration. To recapitulate, by calibration in probability forecasts we mean the extent to which assessed probability is equivalent to proportion correct over a number of assessments of equal probability. For example, if you assign a probability of 0.7 as the likelihood of each of 20 events occurring, 14 of those 20 events should occur *if* your probability forecasts are perfectly calibrated.

Earlier we discussed the usefulness of checking the *consistency* and *coherence* of probability assessments. Perfect test–retest consistency is a necessary but not sufficient condition for perfect coherence, and perfect coherence in turn is a necessary but not sufficient condition for perfect calibration. Consider the case of a sub-standard rule that has been poorly manufactured such that when it reports a measure of one meter it is in fact undermeasuring by one centimeter. In this case, measurements by the rule would be *consistent* from one measurement to another and would also be *coherent* in that a two meter measurement would additively consist of two one-meter measurements. However, the measurement itself is, as we know, invalid. It follows that consistency and coherence are necessary but not sufficient conditions for validity. If assessed subjective probabilities are inconsistent and incoherent they cannot be valid, i.e. well calibrated. The stages in the assessment of probability forecasts are given in Figure 10.3.

Given that assessed probabilities are consistent and coherent then validity becomes an issue. However, *calibration is a long-run measure*. You cannot sensibly compute the calibration of a single probability assessment except for that of 1.0, i.e. certainty. Although perfect calibration is the most desirable aspect of judgmental probability forecasts, in most practical circumstances it may not be possible to measure this aspect of validity. Accordingly, attention should be focused on the other indices of a probability assessment's adequacy; consistency and coherence. As we saw in Chapter 9, our view is that performance-demonstrated expertise in judgmental probability forecasting is underpinned by practice and regular performance feedback. These conditions for learning calibration as a skill are perhaps most obvious in professional weather forecasting. However, as Einhorn[15] has argued, *most* judgmental forecasts are made without the benefit of such accurate feedback. Einhorn traces these difficulties to two main factors. The first is a lack of search for and use of disconfirming evidence, and the second is the use of unaided memory for coding, storing, and retrieving outcome information. In addition, predictions instigate actions to facilitate desired

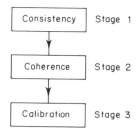

Figure 10.3 Stages in the assessment of probability forecasts

outcomes and indeed, outcome feedback can be irrelevant for correcting poor judgment. Einhorn gives the following example:

> Imagine that you are a waiter in a busy restaurant and because you cannot give good service to all the people at your station, you make a judgment regarding which people will leave good or poor tips. You then give good or bad service depending on your judgment. If the quality of service, in itself, has an effect on the size of the tip, outcome feedback will 'confirm' the predictions ('they looked cheap and left no tip—just as I thought'). The extent of such self-fulfilling prophecies is much greater than we think and represents a considerable obstacle to learning from outcome feedback.

It is clear that such 'treatment effects' where actions in the world can determine subsequent outcomes may be more prevalent in some types of forecasting situations than in others. The implications of this research for decision analysis practice are not clear cut. Most forecasting tasks can be seen to be unlike weather forecasting in that there are actions that the forecaster can take to avoid or facilitate possible futures. As we shall see in Chapter 11, decisions once made are often 'made' to work. It follows that simply keeping a tally of your probability forecasts and the subsequent occurrence or non-occurrence of events may not be helpful in evaluating the calibration of your judgmental forecasting. However, such a tally is more likely to reveal poor validity than simply *reflecting* on the quality of your own judgment. This conclusion is founded on what has come to be termed the 'hindsight bias'. Briefly, Fischhoff[16] has shown that *remembered* predictions of future events 'move' towards the event's occurrence or non-occurrence. For example, a *real* prediction of 0.8 that an event will happen tends to be inflated, in recall, with the knowledge that the event did, in fact, occur. Such a hindsight bias limits the possibility of improving calibration in the light of experience and tends to artificially inflate our confidence in our own judgmental abilities, resulting in what Fischhoff has termed the 'I-knew-it-all-along' effect.

ASSESSING PROBABILITIES FOR VERY RARE EVENTS

Assessment techniques that differ from those we have so far discussed are generally required when probabilities for very rare events have to be assessed. Such events are often of interest because of the disastrous consequences which may be associated with their occurrence. The catastrophic failure of a nuclear power plant, the collapse of a dam or the release of toxic fumes from a chemical factory are obvious examples of this event.

Because of the rarity of such events, there is usually little or no reliable past data which can support a relative frequency approach to the probability assessment, and subjective estimates may be subject to biases which result from the use of the availability heuristic. For example, it may be easy to imagine circumstances which would lead to the event occurring even though these circumstances are extremely improbable. Moreover, as von Winterfeldt and Edwards[17] point out, rare events are newsworthy almost by definition, and widespread reports of their occurrence may have the effect of intensifying the availability bias. Decision makers are also likely to have problems in conceiving the magnitudes involved in the probability assessment. It is difficult to distinguish between probabilities such as 0.0001 and 0.000001, yet the first probability is a hundred times greater than the second.

Obviously, a probability wheel would be of little use in assessing probabilities like these. There are, however, a number of ways in which the problems of assessing very low probabilities can be tackled. Event trees and fault trees allow the problem to be decomposed so that the combinations of factors which may cause the rare event to occur can be identified. Each of the individual factors may have a relatively high (and therefore more easily assessed) probability of occurrence. A log-odds scale allows the individual to discriminate more clearly between very low probabilities. We will examine each of these approaches below.

Event Trees

Event trees are the same as the probability trees which we met in Chapter 3. Figure 10.4 shows a simplified tree for a catastrophic failure at an industrial plant. Each stage of the tree represents a factor which might, in combination with others, lead to the catastrophe. Rather than being asked to perform the difficult task of directly estimating the probability of catastrophic failure, the decision maker is asked instead to assess the probability that each factor will contribute to it. Then, by using the multiplication and addition rules of probability, the overall probability of failure can be calculated (note that, for simplicity, here it has been assumed that all the factors operate independently). Of course, it is important to try to ensure that the tree is as complete as possible so that all possible causes of the event in question are included (see Chapter 5). However, it may be difficult or impossible to assess subjective probabilities for such events as human error, deliberate sabotage, and, by definition, unforseen weaknesses in a system.

Fault Trees

Sometimes it is easier to consider the problem from a different point of view. In contrast to event trees, fault trees start with the failure or accident and

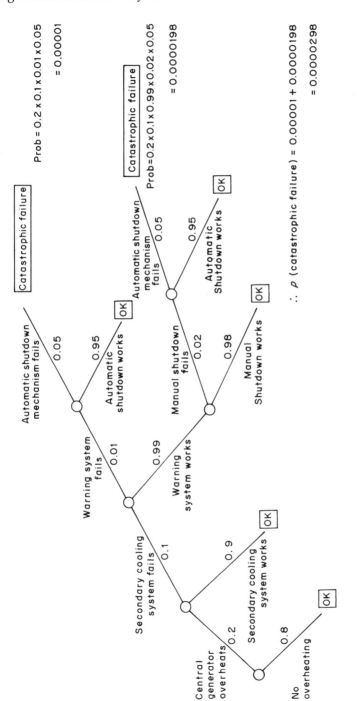

Figure 10.4 An event tree

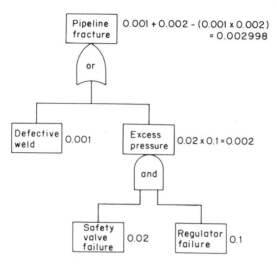

Figure 10.5 A fault tree

then depict the possible causes of that failure. For example, suppose that a decision maker wants to assess the probability that a vital pipeline in a chemical plant will fracture within the next 12 months. Figure 10.4 shows a fault tree for his problem. He considers that a fracture would occur if there is either a defective weld or excess pressure in the pipeline. Because either event on its own would be sufficient to cause the fracture these two events are connected to the 'pipeline fracture' node by an 'or' symbol. The excess pressure in turn, however, would only occur if there was *both* a regulator failure *and* a failure in the safety valve, so an 'and' symbol connects these two events to the 'excess pressure' node.

The decision maker can now assess probabilities for the lowest-level events (these probabilities are shown on the tree) and then work up the tree until he eventually obtains the probability of the pipeline fracturing. Since the safety valve and regulator failures are considered to be independent their probabilities can be multiplied to obtain the probability of excess pressure. This probability can then be added to the probability of a weld being defective to obtain the probability of the pipe fracture occurring. (Note that since excess pressure and a defective weld are not mutually exclusive events the very low probability of them both occurring has been subtracted from the sum—see Chapter 3.) Of course, in practice most fault trees will be more extensive than this one. Indeed, the decision maker in this problem may wish to extend the tree downwards to identify the possible causes of safety valve or regulator failure in order to make a better assessment of these probabilities.

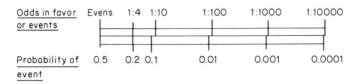

Figure 10.6 A log-odds scale

Using a Log-odds Scale

Because people generally have problems in distinguishing between probabilities like 0.001 and 0.0001 some analysts prefer to use what is known as a log-odds scale to elicit the probabilities of rare events. You will recall that odds represent the probability that an event will occur divided by the probability that it will not. By converting probabilities to odds and then taking the logarithms of the results we arrive at a scale like that shown in Figure 10.6 (this figure shows the scale only for probabilities below 0.5). An analyst using the scale would ask the decision maker to mark the point that represents his or her assessment of the likelihood of the event occurring. Note that the effect of using log-odds is to spread out the ends of the probability scale, making assessment of very high and low probabilities clearer. The scale also ranges from minus to plus infinity, which makes it very difficult for individuals to assert that particular events are impossible or certain to occur. According to Bunn and Thomas,[6] the log-odds scale appears to correspond to the internal scale which individuals use when assessing odds.

SUMMARY

In this chapter we have described the process by which a decision analyst elicits subjective probability assessments from the decision maker. Within the analyst's toolbox of techniques are indirect and direct methods for both discrete and continuous assessments. Consistency and coherence checks are used extensively to police the assessment process, since validity is a more problematic criterion. The chapter concluded with a description of assessment methodologies that are used for very rare events.

EXERCISES

(1) Use the probability wheel and direct estimate methods to elicit a colleague's subjective probabilities for the outcomes of a forthcoming sports event (e.g. a snooker tournament). Check for consistency between

the assessment methods and use the addition rule to evaluate the coherence of the probabilities that your colleague has assessed. Keep a tally of the future outcomes as they occur and draw a calibration graph of the validity of the assessed probabilities.

(2) Use the probability method to elicit a colleague's subjective probability distribution for his or her marks in their next decision analysis assignment.

(3) Repeat the above exercise for predictions of the starting salary distribution in his or her first employment after completing their current educational course of study.

REFERENCES

1. Spetzler, C. S. and Stäel von Holstein, C. A. (1975) Probability Encoding in Decision Analysis, *Management Science*, **22**, 340–352.
2. Lichtenstein, S., Fischhoff, B. and Phillips, L. D. (1981) Calibration of Probabilities: State of the Art to 1980, in D. Kahnemann, P. Slovic and A. Tversky (eds) *Judgment under Uncertainty: Heuristics and Biases*. Cambridge University Press, New York.
3. For more details of the probability approach see Stael von Holstein, C. A. and Matheson, J. (1979) A Manual for Encoding Probability Distributions, *SRI International*, August.
4. Seaver, D. A., Von Winterfeldt, D. and Edwards, W. (1978) Eliciting Subjective Probability Distributions on Continuous Variables, *Organisational Behavior and Human Performance*, **21**, 379–391.
5. Wallsten, T. S. and Budescu, D. V. (1983) Encoding Subjective Probabilities: a Psychological and Psychometric Review, *Management Science*, **29**, 151–173.
6. Bunn, D. W. and Thomas, H. (1975) Assessing Subjective Probability in Decision Analysis, in D. J. White, and K. C. Bowen (eds) *The Role and Effectiveness of Theories of Decision in Practice*, Hodder and Stoughton, London.
7. For example, Beach, L. R. and Phillips, L. D. (1967) Subjective Probabilities Inferred from Estimate and Bets, *Journal of Experimental Psychology*, **75**, 354–359.
8. For example, Winkler R. L. (1967) The Assessment of Prior Distributions in Bayesian Analysis, *Journal of the American Statistical Association*, **62**, 776–800.
9. Phillips, L. D., Hays, W. L. and Edwards, W. (1966) Conservatism in Complex Probabilistic Inference, *IEEE Transactions in Human Factors in Electronics*, **7**, 7–18.
10. Marks, D. F. and Clarkson, J. K. (1972) An Explanation of Conservatism in the Book-bag-and-poker-ship Situation, *Acta Psychologica*, **36**, 245–260.
11. Alberoni, F. (1962) Contribution to the Study of Subjective Probability, *Journal of General Psychology*, **66**, 241–264.
12. Wright, G. and Whalley, P. C. (1983) The Supra-additivity of Subjective Probability, in B. Stigum and F. Wenstop (eds) *Foundation of Risk and Utility Theory with Applications*, Reidel, Dordrecht.
13. Tversky, A. and Kahneman, D. (1974) Judgment under Uncertainty: Heuristics and Biases, *Science*, **185**, 1124–1131.

14. Lindley, D. V., Tversky, A. and Brown, R. V. (1979) On the Reconciliation of Probability Assessments, *Journal of the Royal Statistical Society*, **142**, 146–180. See also Ayton, P. and Wright, G. (1987) Assessing and Improving Judgmental Probability Forecasts, *Omega*, **15**, 191–196, and Wright, G., Ayton, P. and Whalley, P. C. (1985) A General Purpose Aid to Judgmental Probability Forecasting, *Decision Support Systems*, **1**, 333–340.

15. Einhorn, H. J. (1980) Overconfidence in Judgment. In Shweder, R. A. and Fiske, D. W. (eds) *New Directions for Methodology of Social and Behavioral Science: No 4*, Jossey-Bass, San Francisco.

16. Fischhoff, B. (1975) Hindsight = Foresight: The Effect of Outcome Knowledge on Judgment under Uncertainty, *Journal of Experimental Psychology: Human Perception and Performance*, **1**, 288–299.

17. Von Winterfeldt, D. and Edwards, W. (1986) *Decision Analysis and Behavioral Research*, Cambridge University Press, New York.

Chapter 11

Decisions Involving Groups of Individuals

INTRODUCTION

This book has spent many chapters looking at the use of decision analysis for improving decision making. As we have seen, the major inputs to the analysis are subjective probabilities, utilities and decision tree structures. So far, we have focused on the individual decision maker, but important decisions are often made by accountable managers working within small groups of people, most, or all, of whom have information that could be utilized in the decision-making process. Often individuals may differ in their subjective probabilities of events, their utilities of outcomes or in their perception of the subsequent actions available as the pattern of actions, events and outcomes are unfolded into the future.

If the opinion and values of individuals differ, how should the differences be resolved? Obviously, several individuals who are involved in decision making bring together a larger fund of experience, knowledge and creative insights. It is intuitively reasonable that the chances of overlooking possible events and possible courses of action are diminished in group decision making. Indeed, the synergy of individuals may make the overall quality of the group decision greater than the sum of the parts. The creation of juries, panels and cabinets as ways of reaching decisions can be seen to be based on this premise.

This chapter describes and evaluates ways of combining individual judgments to produce 'improved' judgments. There are essentially two approaches to the problem: mathematical and behavioral aggregation (although the approaches can be combined). Mathematical aggregation, which we will discuss first, involves techniques such as the calculation of a simple average of the judgments of the individual group members. In behavioral aggregation a group judgment is reached by members of the group communicating with each other either in open discussion or via a more structured communication process.

Two simple advantages arise from obtaining group judgments in decision analysis. First, more information about possible ranges of utilities and probabilities can be obtained, and it is then possible to perform sensitivity analysis on these ranges to see if the decision specified by the analysis is changed by these variations. Second, a group of people who are involved in such a decision process may become more committed to implementing the decision which is eventually made. As we shall see in the section on decision conferencing, this latter advantage can be a major one.

MATHEMATICAL AGGREGATION

Ferrell[1] provides an excellent and comprehensive discussion of mathematical and other aggregation methods. Much of the following discussion is based on his review.

There are a number of advantages to be gained by using mathematical aggregation to combine the judgments of the individual members of a group. In particular, the methods involved are relatively straightforward. For example, we might ask each member of a group to estimate the probability that the sales of a product will exceed 10 000 units next year and then calculate a simple average of their estimates. This means that the more complex and time-consuming procedures of behavioral aggregation are avoided. Moreover, the group members do not have to meet. Their judgments can be elicited by telephone, post or computer and therefore the influence of dominant group members is avoided. However, there can be serious problems with the mathematical approach as the following, rather contrived, example shows.

Suppose that a production manager and an accountant have to make a joint decision between investing in a large- or small-volume processor. The payoff of the processor will depend upon the average level of sales which will be achieved during its useful life. Table 11.1 shows the production manager's subjective probabilities for the sales levels and his utilities for the different actions and outcomes. It can be seen that the expected utility of the high-volume processor is 0.4 (i.e. $0.4 \times 1 + 0.6 \times 0$) while for the low-volume processor it is 0.412 (i.e. $0.4 \times 0.1 + 0.6 \times 0.62$), so the production manager will just prefer the low-volume processor.

Table 11.2 shows the accountant's view of the problem. Her expected utilities are 0.5 for the high-volume processor and 0.51 for the low-volume one so she will also favor the low-volume processor.

Table 11.1 The production manager's utilities and probabilities

Action	Average sales levels		(Utilities)
	High	Low	Expected utility
Buy high-volume processor	1.0	0	0.4
Buy low-volume processor	0.1	0.62	0.412
Probabilities	0.4	0.6	

Table 11.2 The accountant's utilities and probabilities

Action	Average sales levels		(Utilities)
	High	Low	Expected utility
Buy high-volume processor	1.0	0	0.5
Buy low-volume processor	0.52	0.5	0.51
Probabilities	0.5	0.5	

Table 11.3 The average of the utilities and probabilities

Action	Average sales levels		(Utilities)
	High	Low	Expected utility
Buy high-volume processor	1.0	0	0.45
Buy low-volume processor	0.31	0.56	0.4475
Probabilities	0.45	0.55	

However, if we now take the average of the probabilities and utilities of the two individuals, we arrive at the figures in Table 11.3. If these figures are used to make the decision it can be seen that the 'preferred' group choice is the high-volume processor, despite the fact that both individuals prefer the low-volume one! We will discuss later whether it is valid or meaningful to average subjective probabilities or utilities, but first let us consider methods which can be used to aggregate judgments in general.

AGGREGATING JUDGMENTS IN GENERAL

Single-value estimates of factors such as costs, sales or times to complete a project are often used in decision analysis models when the use of a probability distribution for every unknown quantity would lead to a model which was too complex to be useful. Two methods of combining individual estimates of unknown quantities are considered below.

Taking a Simple Average of the Individual Judgments

First, let us examine the situation where the individual group judgments can be regarded as being unbiased (i.e. there is no tendency to over- or underestimate), with each person's estimate being equal to the true value plus a random error which is independent of the errors of the other estimates. In these circumstances it can be shown that taking the simple average of the individual estimates is the best way of aggregating the judgments. The reliability of this group average will improve as the group size increases because the random error inherent in each judgment will be 'averaged out'. However, each additional member of the group will bring progressively smaller improvements in reliability, so that a point will be reached where it will not be worth the effort or cost of extending the group because a sufficiently reliable estimate can be achieved with the existing membership.

The situation described above is rarely encountered in practice. Generally, the members of the group will produce estimates which are positively correlated. For example, if one member has overestimated next year's sales there will be a tendency for the other members to do likewise. This is likely to occur because group members often have similar areas of expertise or because they all work in the same environment where they are exposed to the same sources of information. If there is a high intercorrelation between the judgments of the group members, then little new information will be added by each additional member of the group and there may be little point in averaging the judgments of more than a small group of individuals. For example, Ashton and Ashton[2] conducted a study in which a group of 13 advertising personnel at *Time* magazine were asked to produce forecasts of the number of advertising pages that would be sold by the magazine annually. When the simple average of individuals' forecasts was used, it was found that there was little to be gained in accuracy from averaging the forecasts of more than five individuals.

Taking a Weighted Average of the Individual Judgments

When some members of the group are considered to be better judges than others then it may be worth attaching a higher weight to their estimates and using a weighted average to represent the group judgment. For example, suppose that three individuals, Allen, Bailey and Crossman, make the following estimates of the cost of launching a new product: $5 million, $2.5 million and $3 million. We think that Allen is the best judge and Crossman the worst, and we therefore decide to attach weights of 0.6, 0.3 and 0.1 to their estimates (note that if the weights do not sum to one then this can be achieved by normalizing them—see Chapter 2). The group estimate will therefore be:

$$(0.6 \times \$5m) + (0.3 \times \$2.5m) + (0.1 \times \$3m) = \$4.05m$$

Clearly, the main problem of using weighted averages is that the judgmental skills of the group members need to be assessed in order to obtain the weights. Methods which have been proposed fall into three categories: self-rating, rating of each individual by the whole group (see, for example, De Groot[3]) and rating based on past performance. However, there can be difficulties in applying these methods. The first two approaches compound the individual's judgmental task by requiring not only judgments about the problem in hand but also those about the skill of individual group members. In some circumstances these problems can be avoided by using weights based on past performance, but as Lock[4] points out, even here there can be difficulties. The current judgmental task may not be the same as those in the past. For example, the quantity which an individual has to judge may be less familiar than those which have been estimated previously. Furthermore, past performance may be a poor guide where judges have improved their performance through learning.

Clearly, simple averaging avoids all these problems, so is it worth going to the trouble of assessing weights? Research in this area has consistently indicated that simple averages produce estimates which are either as good as, or only slightly inferior, to weighted averages (see, for example, Ashton and Ashton[2]). Ferrell[1] suggests a number of reasons for this. He argues that groups tend to be made up of individuals who have very similar levels of expertise and access to the same information. In the case of small groups, even if we are fortunate enough to identify the best individual estimate, its accuracy is unlikely to be much better than that of the simple average of the entire group's judgments. Ferrell also points out that few experiments provide the conditions where weighting would be likely to offer advantages. He suggests that these conditions exist when there is:

> a moderately large group of well-acquainted individuals that frequently works together and has a wide range of different types of expertise to bring to bear on questions that require an equally wide range of knowledge.

In the absence of these conditions a simple average of individuals' judgments will probably suffice.

AGGREGATING PROBABILITY JUDGMENTS

There are particular problems involved when probabilities need to be aggregated, as the following example shows. The managers of a construction company need to determine the probability that a civil engineering project will be held up by both geological problems and delays in the supply of equipment (the problems are considered to be independent). Two members of the management team are asked to estimate the probabilities and the results are shown in Table 11.4. It can be seen that the 'group' assessment of the probability that both events will occur differs, depending on how the averaging was carried out. If we multiply each manager's probabilities together and then take an average we arrrive at a probability of 0.24. However, if we first average the managers' probabilities for the individual events and then multiply these averages together we obtain a probability of 0.225.

Table 11.4

	p (geological problems)	p (equipment problems)	p (both)
Manager 1's estimates	0.2	0.6	$0.2 \times 0.6 = 0.12$
Manager 2's estimates	0.4	0.9	$0.4 \times 0.9 = 0.36$
			Average = 0.24 But:
Average of the estimates	0.3	0.75	$0.3 \times 0.75 = 0.225$

Because of these types of problem a number of alternative procedures have been suggested for aggregating probabilities. One approach is to regard one group member's probability estimate as information which may cause another member to revise his or her estimate using Bayes' theorem. Some of the methods based on this approach (e.g. Morris[5] and Bordley[6]) also require an assessment to be made of each individual's expertise and all are substantially more complicated than simple averaging.

Another approach is to take a weighted average of individual probabilities, using one of the three methods of weighting which we referred to earlier. However, again there appears to be little evidence that weighting leads to an improvement in performance over simple averaging (see, for example, Winkler[7] and Seaver[8] as cited in Ferrell[1]).

What are the practical implications of this discussion? The most pragmatic approach to aggregating probabilities would appear to be the most straightforward, namely, to take a simple average of individual probabilities. This method may not be ideal, as our example of the civil engineering project showed, but as Von Winterfeldt and Edwards[9] put it: 'The odds seem excellent that, if you do anything more complex, you will simply be wasting your effort.'

AGGREGATING PREFERENCE JUDGMENTS

When a group of individuals have to choose between a number of alternative courses of action is it possible, or indeed meaningful, to mathematically aggregate their preferences to identify the option which is preferred by the group? To try to answer this we will first consider decision problems where the group members state their preferences for the alternatives in terms of simple orderings (e.g. 'I prefer A to B and B to C'). Then we will consider situations where a value or a utility function has been elicited from each individual.

Aggregating Preference Orderings

One obvious way of aggregating individual preferences is to use a simple voting system. However, this can result in paradoxical results, as the following example shows.

Three members of a committee, Messrs Edwards, Fletcher and Green, have to agree on the location of a new office. Three locations, A, B and C, are available and the members' preference orderings are shown below (note that > means 'is preferred to'):

Member	Preference ordering
Edwards	A > B > C
Fletcher	B > C > A
Green	C > A > B

If we now ask the members to compare A and B, A will get two votes and B only one. *This implies: A > B.* If we then compare B with C, B will get two votes and C only one, which implies that *B > C.* Finally, if we were to compare A with C, C would get two votes and A only one, which implies *C > A.* So not only do we have A > B > C but we also have C > A, which means that the preferences of the group are not transitive. This result is known as *Condorcet's paradox.*

In many practical problems alternatives are not compared simultaneously, as above, but sequentially. For example, the committee might compare A with B first, eliminate the inferior option and then compare the preferred option with C. Unfortunately, the order of comparison has a direct effect on the option which is chosen as shown below.

If the group compared A with B first then A would survive the first round. If A was then compared with C, C would be the location chosen. Alternatively, if the group compared B with C first, B would survive the first round. If B was then compared with A then location A would be chosen.

Moreover, a clever group member could cheat by being dishonest about his preferences if the preferences of the other members are already known. Suppose that locations A and B are to be compared first. Edwards realizes that this will make C, his least-preferred location, the final choice. He would prefer B to be selected, so he dishonestly states his preferences as $B > A > C$. This ensures that B, not A, will survive the first round and go on to 'defeat' C in the second.

These sorts of problems led Arrow[10] to ask whether there is a satisfactory method for determining group preferences when the preferences of individual members are expressed as orderings. He identified four conditions which he considered that a satisfactory procedure should meet:

(1) The method must produce a transitive group preference order for the options being considered.
(2) If every member of the group prefers one option to another then so must the group. (You will recall that this condition was not fulfilled in the production manager/accountant's problem which we considered earlier.)
(3) The group choice between two options, A and B, depends only upon the preferences of members between *these* options and not on preferences for any other option. (If this is not the case then, as we saw above, an individual can influence the group ordering by lying about his preferences.)
(4) There is no dictator. No individual is able to impose his or her preferences on the group.

In his well-known *Impossibility Theorem* Arrow proved that no aggregation procedure can guarantee to satisfy all four conditions. Not surprisingly, this significant and rather depressing result has attracted much attention over the years. It suggests that it is impossible to derive a truly democratic system for resolving differences of opinion. Any method which is tried will have some shortcoming.

Given that no method can be perfect is it possible to devise an approach which is reasonably acceptable? Ferrell argues that *approval voting* is both simple and robust. In this system individuals vote for all the options which they consider to be at least just acceptable. The group choice will then be the option which receives the most votes. Of course, this method ignores much of the available information about individual preferences. While you may consider alternatives A and B to be acceptable, you may have a preference for A. However, by ignoring this information the method avoids the sort of paradoxical results which we have seen can occur with other methods.

Aggregating Values and Utilities

It is important to note that Arrow's Impossibility Theorem refers only to situations where individuals have stated the order of their preferences. A statement giving an individual's preference order does not tell you about that person's intensity of preference for the alternatives. For example, when considering three possible holiday destinations you may list your preferences from the best to the worst, as follows:

> Rio de Janeiro
> San Francisco
> Toronto

However, your intensity of preference for Rio de Janeiro may be very much greater than that for San Francisco, while your preference for San Francisco may be only slightly greater than that for Toronto. As we saw in earlier chapters, an individual's intensity of preference for an alternative can be measured by determining either the value or, in the case of uncertainty, the utility of that course of action. The problem with aggregating the values or utilities of a group of individuals is that the intensities of preference of the individuals have to be compared. To illustrate this, let us suppose that a group of two people, A and B, have to choose between our three holiday destinations. For each person, values are elicited to measure their relative preference for the destinations. These values are shown below with 100 representing the most preferred destination and 0 the least preferred.

Destination	Person A	Person B	Average
Rio de Janeiro	100	50	75
San Francisco	40	100	70
Toronto	0	0	0

If we take a simple average of values for each destination it can be seen that Rio will be the group choice. However, our calculation assumes that a move from 0 to 100 on one person's value scale represents the same increase in preference as a move from 0 to 100 on the other person's scale. Suppose, for the moment, that we could actually measure and compare the strength of preference of the two people on a common scale. We might find that A is less concerned about a move from his best to his worst location than B, so that if we measure their value scales against our common strength of preference scale we have a situation like that shown in Figure 11.1. The individuals' values measured on this common scale are shown below and it can be seen that San Francisco would now be the group choice.

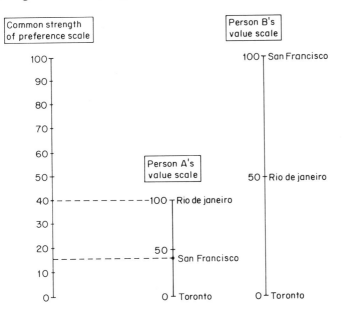

Figure 11.1 Measuring individuals' strengths of preference against a common scale

Destination	Person A	Person B	Average
Rio de Janeiro	40	50	45
San Francisco	16	100	58
Toronto	0	0	0

Is it possible to compare one individual's strength of preference with another's? French[11] examines a number of ways in which such comparisons could be made in theory and shows that all the methods are likely to fail in practice. For example, the whole group might agree unanimously that person 1 prefers P to Q more than person 2 prefers X to Y. But what would happen in the likely event of this unanimity not being achieved? Alternatively, we might use money as the common yardstick for strength of preference. If person A is only prepared to pay $100 to transfer from his worst to his best holiday destination while person B is prepared to pay $1500 then can we say that B's strength of preference is greater than A's? If we answer yes to this question then we must assume that $1 is equally valued by the two people, but what if A is a pauper and B a millionaire? Clearly, our 'objective' scale would lead us back to the problem of comparing individuals' strengths of preference!

In the absence of any obvious method for making interpersonal comparisons of intensity of preference then it seems that our search for a group value or utility function is futile. Nevertheless, the concepts of value and utility can still be useful in group decision making. The derivation of individual values and utilities can help each group member to clarify his or her personal understanding of the problem and also to achieve a greater appreciation of the views of other members. Moreover, a simple average of values and utilities may be useful in providing a rough initial model of the problem. Sensitivity analysis can then be used to test the effect of using individual values and utilities. This may reveal, for example, that certain options are to be preferred to others, irrespective of which individual's utility function is used. At the very least, the process should lead to a more informed discussion and debate.

This reference to group discussion and debate leads us to the next section, where behavioral aggregation methods are considered. We begin by looking at the behavioral problems which can occur when a group of decision makers meet to agree upon a course of action.

UNSTRUCTURED GROUP PROCESSES

One of the major conclusions of research work on descriptions of group decision making is that of well-documented shortcomings. The presence of powerful individuals can inhibit the contribution of those who are lower down the hierarchy. Talkative or extroverted members may dominate the discussions. Indeed, variations in seating arrangements can tend to direct or inhibit individuals' contributions.

Janis[12] has documented a phenomenon that he has termed 'groupthink' within group decision processes. Groupthink is essentially the suppression of ideas that are critical of the 'direction' in which a group is moving. It is reflected in a tendency to concur with the position or views that are perceived to be favored by the group. Of course, such forces may produce speedy judgments and commitment to action. However, such cohesive groups may develop rationalizations for the invulnerability of the group's decision and inhibit the expression of critical ideas. These pitfalls of groupthink are likely to result in an incomplete survey of alternative courses of action or choices. Such incomplete search through the decision space may result in a failure to examine the risks of preferred decisions and a failure to work out contingency plans if the preferred course of action cannot be taken.

Overall, there have been very few laboratory tests of Janis's theory. One main reason is that laboratory researchers have found it difficult to achieve high levels of group cohesiveness, a primary antecedent of groupthink. Another approach to the verification of the theory has been the study of case histories.

One such recent study, by Esser and Lindoerfer,[13] analyzed the decision to launch the space shuttle Challenger on 28 January 1986. The outcome of that flight, the death of all seven crew members within minutes of launch, focused attention on the process leading to the decision to launch. In these researchers' content analysis of the verbal transcripts of a Presidential Commission report on the disaster, statements therein were coded as either positive or negative instances of the observable antecedents and consequences of groupthink. During the 24 hours prior to the launch of the Challenger the ratio of positive to negative items increased significantly. During this time the Level III NASA management were facing increased difficulties in maintaining their flight schedule, and this was expressed as direct pressure on the dissenters who wanted to delay the flight (the engineers) and 'mindguarding'. Mindguarding essentially refers to the removal of doubts and uncertainties in communications to others. In this instance, the Level III NASA management said to the engineers that they would report the engineers' concerns to the Level II NASA management, but they did not.

STRUCTURED GROUP PROCESSES

Awareness of the factors that can degrade group decision making combined with the implicit belief that group judgment can potentially enhance decision making has led to a number of *structured* methods to enhance group decision making by removing or restricting interpersonal interaction and controlling information flow. One such major method has been Delphi.[14] Essentially, Delphi consists of an iterative process for making *quantitative* judgments. The phases of Delphi are:

(1) Panellists provide opinions about the likelihood of future events, or when those events will occur, or what the impact of such event(s) will be. These opinions are often given as responses to questionnaires which are completed individually by members of the panel.
(2) The results of this polling of panellists are then tallied and *statistical* feedback of the whole panel's opinions (e.g. range or medians) are provided to individual panellists before a repolling takes place. At this stage, anonymous discussion (often in written form) may occur so that dissenting opinion is aired.
(3) The output of the Delphi technique is a quantified group 'consensus', which is usually expressed as the median response of the group of panellists.

Without any repolling, simply utilizing the median of a group's opinions on, say, the unit sales of a new product in the first year of production will provide more accuracy than that due to at least 50% of the individual panellists. To see this, consider Figure 11.2.

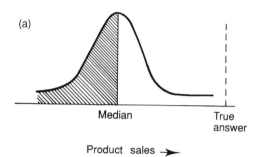

(a)

Median

True
answer

Product sales ➤

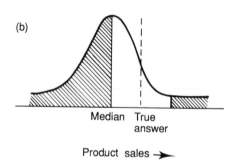

(b)

Median True
answer

Product sales ➤

Figure 11.2

In Figure 11.2(a), where the true answer lies outside the range of estimates, the group median is more accurate than one *half* of the group (the gray shaded area). In Figure 11.2(b), where the true answer lies inside the range of estimates, the group median is more accurate than the *majority* of panellists (the gray shaded areas).

With repolling, and feedback it is assumed that the median response of the group shifts nearer to the true value of the outcome to be predicted. Improvement is thought to result from opinion changes in 'swingers', who change their less firmly grounded opinions, and the opinion stability of 'holdouts', who are assumed to be more accurate than 'swingers'.

Many variations on the theme of Delphi have been tried, including individuals weighting their own expertise to indicate how much they thought that their judgments should contribute to the final tally. Indeed, many procedures for assessing opinions in an anonymous way and then feeding these back to the panel of participants as a basis for re-assessment have been termed Delphi.[25]

However, *experimental* tests of these techniques as ways of improving judgmental accuracy have produced mixed results. They improve performance over the simple average of the individual judgments only slightly and seldom do as well as the *best* member of the group. Ferrell[16] has argued that the reason for the poor performance of the methods is due to the fact that information sharing is small. This is because participants are anonymous and only a simple statistical summary of others' judgments is fed back. It follows that the technique does not help the individual to construct an *alternative theory* or *scenario* with which to produce a revised prediction. Ferrell argues that one may be encouraged to change one's opinion if one's prediction is away from the rest of the group (and one has some uncertainty about the truth of one's answer) and be encouraged to hold tight if one is close to the rest of the group.

In contrast to Delphi techniques, decision conferencing, to be discussed next, presents a socially interactive approach to decision making in order to generate a shared *understanding* of a problem and to produce a commitment to action.

DECISION CONFERENCING

Decision conferencing was invented in the late 1970s by Cameron Peterson at a US-based consulting firm called Decision and Designs Incorporated. Essentially, decision conferencing brings together decision analysis, group processes and information technology over an intensive two- or three-day session attended by people who wish to resolve a complex issue or decision. In this context, a small group of people who have an input to a major decision are often seated on the perimeter of a round table and talk through their problem with a decision analyst, who acts to facilitate group interactions and knowledge sharing. In the background another decision analyst uses interactive decision-aiding technology to model individual and group views on such issues as multi-attribute option evaluation and resource allocation. As we have seen in earlier chapters, the outputs of such modeling seldom agree with unaided holistic judgments. One major responsibility of the decision analyst is to explain the underlying logic of the modeling methods to the decision makers. Only if the decision makers can fully appreciate the methods are they likely to accept model-based choices over their own intuitive judgments. To quote Phillips:[17]

> As the results of the modeling become available to the participants, they compare these results to their holistic judgments. It is the inevitable discrepancies that arise, especially early in the modeling process, that drive the dialectic. By exploring these discrepancies, understanding deepens and

changes, and new perspectives are reflected back as changes to the model.
Eventually, participants are satisfied with the model and unable to derive any further insights from it . . . The model has served its purpose.

Phillips is concerned not to impose an optimal solution by black box methods:

> If exploration of the discrepancy between holistic judgment and model results show the model to be at fault, then the model is not requisite—it is not yet sufficient to solve the problem. The model can only be considered requisite when no new intuitions emerge about the problem . . . Requisite models are not produced from people's heads, they are *generated* through the interaction of problem owners.

Therefore the fundamental objective behind decision conferencing is to provide a synthesis of decision analysis techniques and the positive characteristics and dynamics of small-group decision making. Shared understandings of problems are generated with the aid of decision-analytic techniques and social facilitation. Participants gain a sense of common purpose and a commitment to action. Sensitivity analysis allows participants to see if individual disagreements make a difference in the final preferred alternative or decision. Decision-analytic principles provide a *guide* to action, not a black box prescription for action.

We feel that it is intuitively reasonable that decisions which are conferenced to consensus are more likely to be implemented than the output prescriptions of complex black box decision analyses, which involve but a single decision maker who may well have to justify his or her decision to others in the organization. In addition, decisions made by such groups are likely, because of the group commitment, to be 'made' to work.

However, a major question which still remains to be answered is: Are decisions that are conferenced to consensus more or less valid than unaided judgment or prescriptive solutions? For example, does the situational context of decision conferencing produce conditions for groupthink? Phillips[17] has argued that this is not so, since:

(1) Participants are not on home ground. Often decision conferences take place in hotels or an especially designed room on the decision analyst's premises.
(2) The small group is carefully composed of people representing *all* perspectives on the issue to be resolved so that adversarial processes operate in the group to check bias and explore alternative framings of the decision problem.
(3) The decision analyst who acts to facilitate the conference is a neutral outsider who is sensitive to the unhelpful effects of groupthink and reflects this back to the group.

Recently, McCartt and Rohrbough[18] have addressed the problem of evaluating the effectiveness of decision conferencing. These investigators argued that attempts to link good decision outcomes to particular types of group decision support is extraordinarily difficult, since virtually all real-world applications of group decision support do not provide enough baselines of comparison (e.g. tests of alternative methods/techniques or alternative decisions) to satisfy laboratory-based experimental researchers.

For example, as noted above, with group commitment, poor decisions may be 'made' to produce good outcomes, otherwise the credibility of the senior executives who attended the decision conference would be in doubt. Good judgment and decision making have been seen as one of the major character-istics of good managers! McCartt and Rohrbough conclude that any assessment of the effectiveness of a group decision process must be directed at the *process* itself and not to subsequent outcomes. In their study, these investigators followed up a cross-section of 14 decision conferences held by Decision Techtronics at the State University of New York at Albany. Using mailed questionnaires, they enquired about the perceived organizational benefits in the form of improved information management, planning, efficiency and morale. Those decision conferences which were rated as effective were found to be where participants perceived real benefit in the support of the decision analysis techniques and in the opportunity for open and extended discussion about the models that had been built. Ineffective decision conferences were characterized by executive teams who convened to discuss a problem but felt little pressure to reach consensus or construct a plan of action.

To date, well over 500 decision conferences have been conducted in many countries worldwide. The service is now offered by about 15 organizations and shows every sign of developing into an effective way of helping managers deal with complex issues facing their organizations.

SUMMARY

In this chapter we described the advantages and disadvantages of mathematical and behavioral aggregation of opinion. Within mathematical aggregation techniques, simple and weighted averaging were contrasted. Unstructured and structured behavioral aggregation were described and the processes involved in decision conferencing were outlined. Issues in the evaluation of the usefulness of decision conferencing were introduced.

DISCUSSION QUESTIONS

(1) Is consensus best achieved by committee discussion or by mathematical aggregation?
(2) Are structured methods for the behavioral aggregation of opinion useful in practice?

(3) What are the advantages and disadvantages of the decision conferencing approach to decision analysis as contrasted to a more traditional decision analysis involving a single decision maker?

REFERENCES

1. Ferrell, W. R. (1985) Combining Individual Judgments, in G. Wright (ed.) *Behavioral Decision Making*, Plenum Press, New York, pp. 111–145.
2. Ashton, A. H. and Ashton, R. H. (1985) Aggregating Subjective Forecasts: Some Empirical Results, *Management Science*, **31**, 12, 1499–1508.
3. De Groot, M. H. (1974) Reaching a Consensus, *Journal of the American Statistical Association*, **69**, 118–121.
4. Lock, A. (1987) Integrating Group Judgments in Subjective Forecasts, in G. Wright and P. Ayton (eds) *Judgmental Forecasting*, Wiley, Chichester.
5. Morris, P. A. (1983) An Axiomatic Approach to Expert Resolution, *Management Science*, **29**, 1, 24–32.
6. Bordley, R. F. (1982) The Combination of Forecasts: a Bayesian Approach, *Journal of the Operational Research Society*, **33**, 171–174.
7. Winkler, R. L. (1971) Probabilistic Prediction: Some Experimental Results, *Journal of the American Statistical Association*, **66**, 675–685.
8. Seaver, D. A., Von Winterfeldt, D. and Edwards, W. (1978) Eliciting Subjective Probability Distributions on Continuous Variables, *Organisational Behavior and Human Performance*, **21**, 379–391.
9. Von Winterfeldt, D. and Edwards, W. (1986) *Decision Analysis and Behavioral Research*, Cambridge University Press, New York.
10. Arrow, K. J. (1951) *Social Choice and Individual Values*, Wiley, New York.
11. French, S. (1988) *Decision Theory: An Introduction to the Mathematics of Rationality*, Ellis Horwood, Chichester.
12. Janis, I. R. (1982) *Groupthink* (2nd edn), Houghton Mifflin, Boston, Mass.
13. Esser, J. K. and Lindoerfer, J. S. (1989) Groupthink and the Space Shuttle Challenger Accident: Towards a Quantitative Case Analysis, *Journal of Behavioral Decision Making*, **2**, 167–177. See also: Park, W. W. (1990) A Review of Research on Groupthink, *Journal of Behavioral Decision Making*, **3**, 229–245.
14. See, for recent reviews, Parente, F. J. and Anderson-Parente, J. K. (1987) Delphi Inquiry Systems, in G. Wright and P. Ayton (eds) *Judgmental Forecasting*, Wiley, Chichester. See also: Rowe, G., Wright, G. and Bologer, F. (in press) Delphi: a Re-evaluation of Research and Theory, *Technical Forecasting and Social Change*.
15. Linstone, H. and Turoff, M. (1975) *The Delphi Method: Techniques and Applications*, Addison-Wesley, Wokingham.
16. Ferrell, W. R. (1990) Aggregation of Judgments or Judgment of Aggregations? in N. Moray, W. Rouse and W. R. Ferrell, (eds) *Robotics, Control and Society: Essays in Honor of Thomas B. Sheridan*, Taylor and Francis, New York.
17. Phillips, L. D. (1984) A Theory of Requisite Decision Models, *Acta Psychologica*, **56**, 29–48.
18. McCartt, A. and Rohrbough, J. (1989) Evaluating Group Decision Support System Effectiveness: A Performance Study of Decision Conferencing, *Decision Support Systems*, **5**, 243–253.

Chapter 12

Resource Allocation and Negotiation Problems

INTRODUCTION

In this chapter we will consider how decision analysis models can be applied to two types of problem which usually involve groups of decision makers. First, we will study problems where a limited resource has to be allocated between a number of alternative uses. For example, a group of product managers may have to decide on how next year's advertising budget should be divided between the products for which they are responsible. Similarly, a local police force might have to determine how its available personnel should be allocated between tasks such as crime prevention, traffic policing and the investigation of serious and petty crimes.

As Phillips[1,2] argues, the central purpose of resource allocation models in decision analysis is to resolve what is known as the *commons dilemma*. On the one hand, an organization can decentralize its decision making, allowing each manager to make the best use of the resources which he or she is allocated. While this delegation of responsibility will probably motivate the managers, it is unlikely that the resulting set of independent decisions will lead to an allocation of resources which is best for the organization as a whole. The alternative is to centralize decision making, but this may be demotivating and the resulting allocation will not take into account the local knowledge of the individual managers. The dilemma can be resolved by the managers meeting as a group, possibly in a decision conference, and examining the effect of trading-off resources between their areas of responsibility. Of course, some managers may find that they lose resources as a result of this process, but these losses should be more than compensated by the increased benefits of re-allocating the resources elsewhere. As we shall see, the number of combinations of options that are available in this sort of decision problem can be very large, therefore a computer is normally required to perform the appropriate calculations. One package which has been designed specifically for this purpose is EQUITY, which was developed by the Decision Analysis unit at the London School of Economics (see Barclay[3]). This package has been used to analyze the problem which we will consider in the first part of the chapter and therefore much of the terminology which is associated with EQUITY has been adopted here.

The second application which will be considered in this chapter relates to problems where two parties are involved in negotiations. Typically, these problems concern a number of issues. For example, in industrial relations disputes, negotiations may involve such issues as pay, holidays and length of the working day. By assessing, for each party, the relative importance of these concerns and the values that they would attach to particular outcomes of the negotiations, decision analysis can be used to help the parties to attain a mutually beneficial settlement.

Both these applications rely heavily on the concepts of multi-attribute value analysis, which were covered in Chapter 2, and the reader should make sure that he or she is familiar with these ideas before proceeding. In particular, it should be noted that we have assumed throughout the chapter that the additive value model is appropriate. Though this is likely to be the case for a very wide range of applications, because of the simplicity and robustness of the model, there will, of course, be some circumstances where decision makers' preferences can only be represented by a more complex model.

MODELING RESOURCE ALLOCATION PROBLEMS

An Illustrative Problem

Consider the following problem which relates to a hypothetical English furniture company. At the time of the analysis the company was selling its products through 28 large showrooms which were situated on the edges of cities and towns throughout the country. Following a rapid expansion of sales in the mid 1980s the company's market had been divided into four sales regions, North, West, East and South, and a manager had been made responsible for each. The North sales region, with nine outlets, had accounted for about 30% of national sales in the previous year, but the region had been economically depressed for some time and the immediate prospects for an improvement in the position were bleak. The West region had only three outlets and the company had been facing particularly stiff competition in this region from a rival firm. In the East region there was an even smaller operation with only two outlets, but this was known to be a potential growth area, particularly in the light of the recent electrification of the main railway line to London. To date, the most successful sales area, accounting for 50% of national sales, had been the South. The company had a major operation here with 14 showrooms but, although the market in this region was buoyant, planning regulations meant that there had been a problem in finding suitable sites for the construction of new outlets.

The management of the company were planning their strategy for the next five years and the main problem they faced was that of deciding how the available resources should be allocated between the sales regions. For example, should they reduce the number of outlets in the North and re-allocate the freed resources to the more promising East region? It was resolved that the key managers involved should meet as a group with a facilitator available to structure a decision analysis model of the problem.

Determining the Variables, Resources and Benefits

The first question the group faced was the determination of the *variables* involved in the problem. The term 'variable', in this context, refers to an area to which resources might be directed. Typical variables might be different research and development projects, different functional areas of a business or different product lines.

In this problem the group soon decided that the four sales regions were the variables, but in some problems the identification of the variables is not so easy. The key point to bear in mind is that, apart from the fact that they are competing for resources, the areas should be regarded as separate compartments with a decision in one area having virtually no affect on a decision in another. For example, suppose that a regional passenger transport authority, which is responsible for railways, bus and underground services, has to decide on the allocation of investment capital for the next planning period. If the three different forms of transport were designated as the variables then the required independence may not exist. In an integrated transport system a decision to invest more in the railway system will probably have implications for the bus transport department: greater rail usage may mean that more buses will be required to provide transport to and from railway stations. In this case it may be better to define the variables as different localities within the region or different groups of routes. Statements such as 'If department X chooses that option, then we ought to do this' may reveal a lack of independence.

Having identified the variables, the managers of the furniture company had to think about the nature of the resources which they would be allocating to the different sales regions. Again this proved to be easy, since their main concern was the efficient use of the company's money. Of course, in some problems there may be several resources which are to be allocated such as personnel, equipment and production facilities.

Next, the group was asked to identify the benefits which they hoped would result from the allocation of money between the regions. After some discussion it was agreed that there were three main objectives:

(i) To sustain the profitability of the company in the short term (we will refer to the benefit associated with this objective simply as 'Profit');
(ii) To increase the company's turnover and national market share by expanding the number of outlets (we will call the associated benefit 'Market share');
(iii) To minimize the risk associated with any developments. (Some strategies would offer the benefit of being less risky, and we will refer to this as 'Risk'. Note that we will be using values rather than utilities here even though the decision involves risk—we discussed this issue in Chapter 2.)

Identifying the Possible Strategies for Each Region

The group of managers then considered the strategies which were available in each region and these are summarized in Figure 12.1. For the North sales

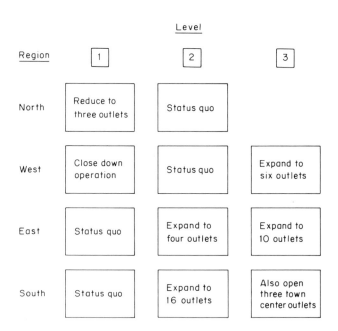

Figure 12.1 Possible strategies identified by managers of the furniture company

region two strategies were identified. Either the operation there could be scaled down so that only three outlets would be retained (this might be sensible in view of the recent poor performance of the company in this region), or the status quo could be maintained. Expanding the number of outlets in the North was not thought to be worth considering.

In the West three strategies were identified. First, it was thought that it might be a good idea to close down the relatively small-scale operation there so that effort could be concentrated elsewhere. Alternatively, either the status quo could be maintained or the number of outlets could be doubled to six. In the promising East region it was thought that the company should either maintain the status quo or expand to either four or 10 outlets.

Finally, in the South either the status quo could be maintained or the company could increase its number of outlets to 16. A third strategy was also considered in the South. In addition to operating 16 outlets, the company could also consider opening three showrooms in town center sites because of the difficulty in finding suitable out-of-town sites in this region.

Note that in Figure 12.1 the strategies for each region are organized in the order of the level of resources which would be required, with level 1 representing the lowest level of expenditure. The figure can also be used to identify the possible combinations of strategies which could be adopted. Each combination of strategies is referred to as a *package*, so that one example of a package would be: reduce the outlets operated in the North to three, expand the West's operation to six outlets and maintain the status quo in the other two regions. Since one of the two strategies in the North could be combined with one of the three strategies in the West and one of the three strategies in the East, and so on, the number of packages that were available was $2 \times 3 \times 3 \times 3$, which equals 54. In many problems this number is much higher. For example, Phillips[1] quotes an application where there were over 100 000 possible packages.

Assessing the Costs and Benefits of Each Strategy

The next stage in the formulation of the model was the assessment of the costs and benefits which were associated with each strategy. Table 12.1 shows the figures which were elicited from the group after much discussion. Before we outline how these figures were arrived at it is important to note that, with the exception of the monetary values, all the values in the analysis are measured on interval scales. The reader will recall from Chapter 2 that this means that we can only compare increases or decreases on the scale, so we will be constantly thinking about such questions as: 'How does the increase in benefits which would result if we switched to strategy X compare with the increase we would achieve if we switched instead to strategy Y?'

Table 12.1 Values of the strategies in the individual regions

Strategies	Costs ($m)	Profits ($m)	Benefits Market share	Risk
NORTH REGION				
(1) Reduce to three outlets	12	(6) 100	0	100
(2) Status quo	28	(−3) 0	100	0
WEST REGION				
(1) Close down operation	−14	(4) 100	0	100
(2) Status quo	7	(−2) 54	30	60
(3) Expand to six outlets	16	(−9) 0	100	0
EAST REGION				
(1) Status quo	2	(3) 100	0	100
(2) Expand to four outlets	25	(0) 91	25	70
(3) Expand to 10 outlets	40	(−30) 0	100	0
SOUTH REGION				
(1) Status quo	20	(65) 100	0	100
(2) Expand to 16 outlets	25	(50) 50	80	40
(3) Add three town center outlets	45	(35) 0	100	0

The first figures elicited from the group were the costs. These represent the estimated amount of money which would need to be spent over the next 5 years in order to carry out each strategy. Note that closing down the operation in the West would be expected to save about $14 million.

Next, the desirability of using each of the strategies was assessed in relation to each benefit, starting with the North region. To do this, a value scale which ranged from 0 to 100 was used with 0 representing the least desirable effect and 100 the most desirable. This assessment was carried out separately by the group for each individual region, so that in Table 12.1 the 100 in the 'Market Share' column in the North region, for example, means that strategy 2 was thought to be the better of the two strategies available in the North for improving market share.

In the case of short-term profits, the group first estimated the profits which might result from the different strategies as monetary values (these figures are bracketed in the table) and these were then converted to values under the assumption that a linear value function was appropriate. Thus in the West region, for example, expanding to six outlets (strategy 3) would result in the least-preferred short-term profit while closing down the operation (strategy 1) was thought to lead to the most desirable profit in the short term. The improvement in short-term profit which would result from switching from strategy 3 to strategy 2 was 54% (the exact figure is 53.8%), as desirable as a switch from strategy 3 to strategy 1.

The other values in the table can be interpreted in a similar way. Of course, in the case of 'Risk' a value of 100 denotes the strategy which was judged to be the least risky. Thus in the West region a switch from strategy 3 to strategy 2 would lead to a reduction in risk which was considered to be 60% as attractive as the reduction in risk which would result if the switch was made from strategy 3 to strategy 1.

Measuring Each Benefit on a Common Scale

Because the values were assessed separately for each region, a movement from 0 to 100 for a particular benefit in one region might be more or less preferable than the same movement in another region. To take this into account, it was now necessary to measure these changes in benefit on a common scale. To illustrate the nature of the problem, let us suppose for the moment that just two regions were operated by the company, West and East, and that a swing from the worst to the best strategy for market share in the West was only seen as half as important as the swing which could be achieved by changing from the worst to the best strategy in the East. Figure 12.2 shows the two value scales for market share side by side. Normally, the longest scale is used as the common scale. Therefore, in this case, if we used the East's value scale as the common scale it can be seen

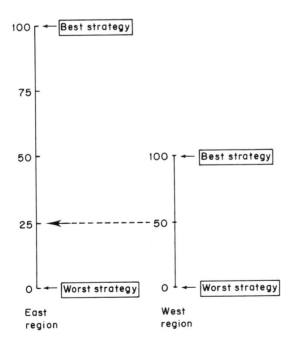

Figure 12.2 A comparison of the East and West regions' value scales for market share

that a value of 50 on the West's scale will only have a value of 25 on the common scale.

The different lengths of the scales are measured by the *within-criterion weights*. In this simple example the East would be allocated a within-criterion weight of 100 for market share and the West a value of 50, which means that on the common scale the West's market share values will only have 50% of their original values. The within-criterion weights that were elicited for the furniture company problem are shown in Figure 12.3 and we next describe how these weights were assessed.

For short-term profit the weights were calculated directly from the group's original monetary estimates. Thus the largest swing between the worst and best short-term profits offered by the different strategies was in the East (a difference of $33 million) and this swing was allocated a weight of 100. The

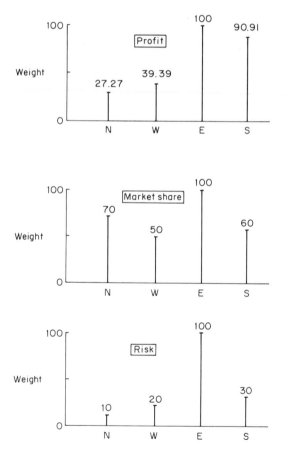

Figure 12.3 The within-criterion weights for the furniture company problem

swing between the worst and best level of profits in the North was only $9 million. Since this was only 27.3% of the largest swing (i.e. $9/33 \times 100$) the within-criterion weight allocated to short-term profits in the North was 27.3. The other profit weights were calculated in a similar way.

For the other benefits the within-criterion weights were elicited using a different approach. In the case of 'Market share', for example, the group was asked to imagine the package which would have the least desirable effect on market share. This would be the package which involved reducing the operation in the North to three outlets, closing down the operation in the West and maintaining the status quo in the other regions. They were then asked: if they could choose to change just one of the strategies in this package, which change would lead to the greatest improvement in market share? After some debate, the group thought that this improvement would be achieved if a switch was made in the East region from maintaining the status quo (strategy 1) to expanding to 10 outlets (strategy 3). This meant that 'Market share' in the East region was assigned a weight of 100.

The facilitator then asked the group to identify, from the strategies available in the other regions, the change which would lead to the second largest improvement in market share. The group said that they would switch from reducing the operation in the North (strategy 1) to maintaining the status quo in that region (strategy 2). After further discussion, they said that they felt that the improvement in market share resulting from this switch would only be 70% of that which would be achieved by the switch they had previously identified in the East region. Hence 'Market share' in the North was allocated a within-criterion weight of 70. The weights for South and West regions were assessed in the same way. A similar approach was used to elicit the within-criterion weights for 'Risk'.

Each benefit now had a common scale which enabled the effect on that benefit of choosing a particular package to be measured. Table 12.2 shows the values of the various strategies measured on these common scales.

Comparing the Relative Importance of the Benefits

Obviously, the managers wanted to be able to assess the overall benefit of using a particular package by combining the values for all three benefits. This meant that they now had to determine a set of weights which would allow one benefit to be compared with the others. These weights could be obtained by directly assessing the relative importance of each benefit, but, as we pointed out in Chapter 2, a better method is to compare the importance of a change (or swing) from the worst position to the best position on one benefit scale with a similar swing on each of the other benefit scales. The resulting weights are known as the *across-criteria weights*.

Table 12.2 Values of strategies with each benefit measured on common scale

Strategies	Costs ($m)	Benefits		
		Profits	Market share	Risk
NORTH REGION				
(1) Reduce to three outlets	12	27.27	0	10
(2) Status quo	28	0	70	0
WEST REGION				
(1) Close down operation	−14	39.39	0	20
(2) Status quo	7	21.27	15	12
(3) Expand to six outlets	16	0	50	0
EAST REGION				
(1) Status quo	2	100	0	100
(2) Expand to four outlets	25	91	25	70
(3) Expand to 10 outlets	40	0	100	0
SOUTH REGION				
(1) Status quo	20	90.91	0	30
(2) Expand to 16 outlets	25	45.46	48	12
(3) Add three town center outlets	45	0	60	0

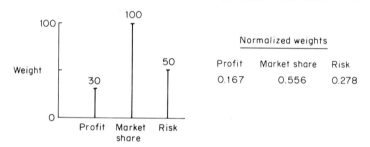

Figure 12.4 The across-criteria weights for the furniture company problem

To derive the weights (which are shown in Figure 12.4) the facilitator looked for a region where a benefit had a within-criterion weight of 100, since this would show where there was the largest swing from the best to the worst position for that benefit. In fact, in this case all three benefits have their biggest swing in the East region (see Figure 12.3). He therefore asked the group to consider this region and asked them to imagine a strategy which offered the worst short-term profit (−$30 million), the poorest prospect for expanding market share and the highest (i.e. least desirable) level of risk. The managers were then asked: if they could change just one of these benefits to its best possible value, which would they choose? The group were unanimous that they would be most concerned to move to the best possible value for market share. This benefit was therefore given an across-criteria weight of 100.

The group's second choice was to move to the best possible position for risk. In fact, a move from the most risky to the least risky position was regarded as only 50% as important as the swing from the worst to the best market share position. 'Risk' was therefore allocated a weight of 50. A move from the lowest to the highest short-term profit (i.e. a move from −$30 million to $3 million) was the least preferred out of the possible swings and, after some discussion, this benefit was assigned a weight of 30. The three across-criteria weights were then normalized (see Chapter 2) and these normalized weights are also shown in Figure 12.4.

It is, of course, always a good idea to apply consistency checks to a group's responses. For example, suppose that the managers had been asked to consider a move from strategy 2 to strategy 3 in the South region. How would the increased market share which would result from this change (an increase of 12 points on the market share value scale—see Table 12.2) have compared with the less attractive level of risk (which results in a loss of 12 points on the risk value scale)? To be consistent, the group should have indicated that the increased market share benefits would be roughly twice as important as the loss in benefits caused by the less attractive level of risk. This is because the across-criteria weights suggested that the market share scale was twice as long as the risk scale.

Identifying the Costs and Benefits of the Packages

It was now possible to identify the overall benefits and costs of any of the packages. At this point the group were asked to propose a package which they felt would lead to the best use of the company's funds. (It was thought that around $70–80 million would be available to support the company's strategies in the four regions.) The package that they suggested was a fairly cautious one. It simply involved maintaining the status quo in every region except the East, where an expansion of operations to four outlets would take place. From Table 12.2 it can be seen that this package would cost $80 million (i.e. $28 million + $7 million + $25 million + $20 million). It would result in profit benefits which would have a total value of 203.2 (i.e. 0 + 21.27 + 91 + 90.91), market share benefits of 110 and risk benefits of 112. Now that the across-criteria weights had been elicited, the overall benefits could be calculated by taking a weighted average of the individual benefits (using the normalized weights) as shown below:

Value of benefits
$$= 0.167 \times (\text{value for profit}) + 0.556 \times (\text{value for market share}) + 0.278 \times (\text{value for risk})$$
$$= (0.167 \times 203.2) + (0.556 \times 110) + (0.278 \times 112)$$
$$= 126.2$$

If similar calculations were carried out for the least beneficial package (as identified by a computer) the value of benefits would be found to be 87.49. The corresponding figure for the most beneficial package is 159.9. The results of the analysis are easier to interpret if these values are rescaled, so that the worst and best packages have values of 0 and 100, respectively. Since 126.2 is about 53.4% of the way between 87.49 and 159.8, the benefits of this package would achieve a value of 53.4 on the 0–100 scale (i.e. this package would give 53.4% of the improvement in benefits which could be achieved by moving from the worst to the best package).

A computer can be used to perform similar calculations for all the other packages and the results can be displayed on a graph such as Figure 12.5. On this graph the efficient frontier links those packages which offer the highest value of benefits for a given cost (or the lowest costs for a given level of benefits).

Note, however, that the packages marked 1 and 2 on the graph do not appear on the efficient frontier, despite the fact that they offer the highest benefits for their respective costs. To see why this is the case, let us consider the choice between the B package and packages 2 and 3 on the graph. Suppose that it has been established that the decision makers think that each extra 'point' gained on the benefit scale is worth $x. This would mean that the value of each package could be measured in monetary terms. For

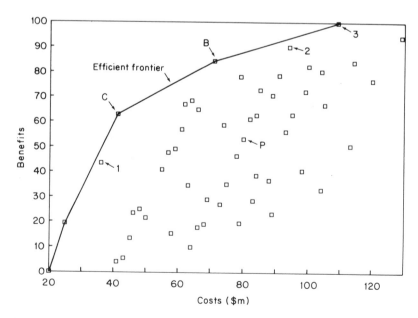

Figure 12.5 Identifying the efficient frontier for the furniture company problem

example, if x is \$2 million the monetary value of the B package would be: (value of benefits) × \$2 million − (cost of package). This equals 84.7 × \$2 million − \$71 million or \$98.4 million. Similarly, it can be shown that the monetary worth of packages 2 and 3 would be \$86.6 million and \$91 million, respectively, so the B package would have the highest monetary worth. Now it can be proved that, whatever monetary value the decision makers were willing to attach to a 'point' on the benefit scale, then either the B package or package 3 would have the highest worth (assuming, of course, that this monetary value was not negative). The fact that package 1 (and, for similar reasons, package 2) would never be preferred accounts for the convex shape of the efficient frontier curve.

It can be seen from Figure 12.5 that the proposed package (represented by P) did not lie on the efficient frontier. When this is the case, the EQUITY package highlights two alternative packages. Package B is a package which will offer the 'best' level of benefits for a cost which is close to that of the proposed package, while package C is a package which will offer roughly the same level of benefits as the proposed package, but at the 'cheapest' level of costs.

Not surprisingly, the group of managers were interested in finding out about package B and the EQUITY program revealed that this involved the following strategies:

In the North: Maintain the status quo
In the West: Expand to six outlets
In the East: Maintain the status quo
In the South: Expand to 16 outlets

This package would cost \$71 million, which was less than the proposed package but would lead to benefits which had a value of 84.7, which were considerably higher. The group were surprised that the package did not involve any expansion in the promising East region.

The explanation for this was partly provided by the EQUITY program which enabled the costs and benefits of each strategy to be compared for individual regions. Figure 12.6 shows the results for the West, East and South regions with values of 100 and 0 representing the highest and lowest levels of benefits *for that region* (the North had only two available strategies so the results for this region were not analyzed). When a graph has a shape like that shown in Figures 12.6(a) or (b) the 'middle' strategy will never be recommended by the computer. The reasons for this are analogous to those which led to the exclusion of packages 1 and 2 from the efficient frontier. This means that in the East the choice is between the status quo and a major expansion to 10 outlets. The

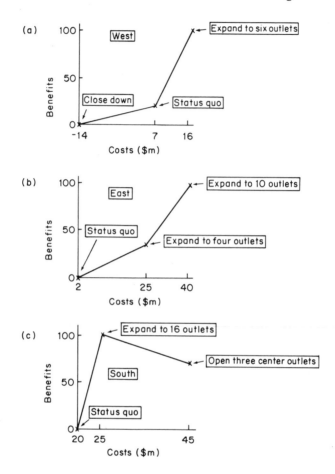

Figure 12.6 Investigating the costs and benefits of strategies in the individual regions

problem is that this major expansion would cost an extra $38 million, and the model suggests that this expenditure is not worthwhile given the limited funds which are available. It can also be seen that the opening of 3 town center sites in the South would actually lead to a loss of benefits, despite an extra expenditure of $20 million. (Note that these graphs do not take into account the different lengths of the regional benefit scales, i.e. the within-criterion weights, so direct comparisons cannot be made between them.)

The group were now seriously considering the B package, but the facilitator suggested that they should first perform a sensitivity analysis before making a final decision.

Sensitivity Analysis

The group's proposed package had been based on the assumption that funds of about $70–80 million would be available to finance the company's strategies. Because there was some uncertainty about how much money would actually be available, the group felt that it might be worth identifying the best packages if the company had more funds at its disposal. For example, it might be that an argument could be made for extra money if it could be shown that this would lead to a package with substantially increased benefits.

The EQUITY package enables the user to explore the packages on the efficient frontier curve. Clearly, if more money is available then this will involve a move along the curve to the right. EQUITY showed that the next package on the curve involved the same strategies as the B package, with the exception that in the East region an expansion to 10 outlets would be recommended. This would lead to the maximum possible value of benefits of 100, but would involve expenditure of $109 million. The group agreed that this level of expenditure would not be feasible and, anyway, the relatively small improvement in benefits over the B package would not have justified the increased costs, even if such funds were available.

The group also investigated the effect of reducing the money which was available to be invested (i.e. the effect of moving to the left of the B package on the efficient frontier curve). The computer showed that the next package on the curve would cost only $41 million, but would only generate benefits which had a value of 63 (in fact this is the C package). There was little enthusiasm for this package which would have involved the same strategies as the B package in all regions except the West, where the company's operations would be closed down completely.

One of the group next suggested that not enough attention had been paid in the model to risk. Rather than allocating an across-criteria weight of 50, she felt that a value of 70 would have been more appropriate. However, when this new value was entered into the computer the program showed that there was no change in the strategies recommended by the B package. This shows again that models are often quite robust to changes in the figures elicited from decision makers, so there is no need to be concerned about whether the assessments are perfectly accurate. Moreover, although members of the group may disagree about weights, these differences often do not matter and, if this is the case, there is little point in spending time in debating them. When the divergent views of the group members would lead to substantially different packages then, clearly, these differences need to be explored. However, when this does happen the modeling process should, at the very least, have the effect of increasing each member's understanding of how the other group members perceive the problem, and this should lead to more informed debate.

SUMMARY OF THE MAIN STAGES OF THE ANALYSIS

It can be seen that the decision analysis approach to resource allocation problems involves a number of stages, which we have summarized below to give an overview of the process. It is important to note, however, that the analysis rarely proceeds smoothly from one stage to the next. As a greater understanding of the problem emerges it will often be found necessary to return to earlier stages in order to make revisions to the original model.

Stage 1: Identify the resources which are to be allocated, the areas (variables) to which they can be allocated and the various benefits which it is hoped this allocation will achieve.

Stage 2: Identify the possible strategies which are available for each variable.

Stage 3: For each variable, assess the costs and benefits of the different strategies.

Stage 4: Assess the within-criterion weights so that each benefit can be measured on a common scale.

Stage 5: Assess the across-criteria weights which will enable the values for the different benefits to be combined on an overall benefit scale.

Stage 6: Use a computer to calculate the costs and benefits for every package and identify the efficient frontier.

Stage 7: Propose a package which appears to achieve the desired objectives within the constraints of the resources available.

Stage 8: Use the computer to find if there are packages on the efficient frontier which offer higher benefits for the same cost as the proposed package (or the same benefits at a lower cost).

Stage 9: Perform a sensitivity analysis on the results to identify which other packages would be worth considering if more or less resources are available and to examine the effects of changes in the data used in the model.

NEGOTIATION MODELS

Having considered the role of decision analysis in problems where a group of people have to agree on the allocation of scarce resources we now turn our attention to situations where individuals or groups of decision makers find themselves involved in disputes which need to be resolved by negotiation. As Raiffa[4] points out in his highly readable book *The Art and Science of Negotiation*, negotiations can be characterized in a number of ways. For example, they may involve two (or more than two) parties and these parties may or may not be monolithic in the sense that within each party there may be several different interest groups. Some negotiations involve just one issue (e.g. the price at which a house is to be sold), while in other cases several issues need to be resolved (e.g. the weekly pay, holidays and training to which an employee will be entitled). Also, factors such as whether or not there are time constraints, whether or not the final agreement is binding, the possibility (or otherwise) of third-party intervention and the behavior of the participants (e.g. are they honest, have they used threats?) will all vary from one negotiation problem to another.

Of course, some of the techniques we have met in earlier chapters might be useful when decision makers are involved in negotiations. For example, decision trees can be used to represent the options open to a negotiator and the possible responses of his opponent. Our intention here is to focus on an approach which can be helpful when a dispute involves just *two* negotiating parties who would like to reach agreement on *several* issues. As we shall see, it is possible to exploit the different levels of importance which the parties attach to each issue in order to achieve joint gains and so reach deals which are beneficial to both parties.

AN ILLUSTRATIVE PROBLEM

The management of a hypothetical engineering company were engaged in negotiations with a trade union. The union had put forward a package in which they demanded a 15% pay rise, an extra 3 days' holiday per year for all employees and the reinstatement of a group of workers who were fired after committing a breach of company regulations earlier in the year.

Figure 12.7(a) show the management's value function over the range of possible pay awards (it was thought that an award of at least 3% would have to be conceded), together with their estimate of what the union's function would look like. Obviously, management attached the lowest value to an award of 15%, while this was the award which was most preferred by the union. Figures 12.7(b) and (c) show similar curves for the other two issues.

Weights were then elicited from the management team to reflect their view of the relative importance of swings from the worst to the best position on each issue, and these weights are also shown in Figure 12.7. Thus for the management team a swing from a 15% award to a 3% award was seen as the most important: a move from granting 3 days' holiday to no days was only 50% as important as this and agreeing to the union's demand for worker reinstatement was only 10% as important. These weights, which were subsequently normalized, enabled the overall value to management of a particular deal to be measured. The management then made an assessment of the weights they thought that the union would place on similar swings and these are also shown in Figure 12.7.

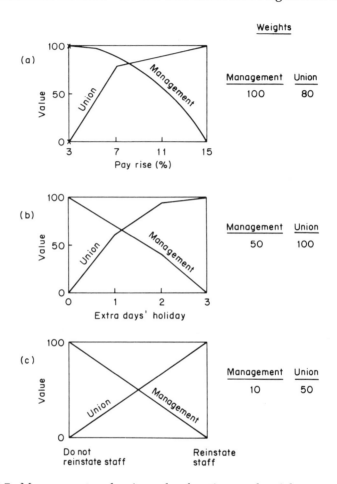

Figure 12.7 Management and union value functions and weights

The values to the management and union of any deal could now be calculated. After several meetings, the following tentative agreement had been reached: only a 3% pay award would be granted but employees would receive an extra 3 days' holiday per year and the sacked workers would be reinstated. Table 12.3 shows how the value of this deal to the management and the union was calculated.

During the negotiations one of the management team had used a computer to calculate the management and union values for all possible deals (assuming, for simplicity, that the percentage increase in pay would be a whole number). The results of these calculations are shown in Figure 12.8. On this graph the efficient frontier represents the set of agreements which are such that no party can improve its position without worsening the

Table 12.3 Calculation of values for the tentative management–union deal

	Value	Normalized weight	Value × weight
Management			
Pay rise of 3%	100	0.625	62.5
3 extra days' holiday	0	0.3125	0
Reinstatement of staff	0	0.0625	0
		Value of deal	62.5
Union			
Pay rise of 3%	0	0.348	0
3 extra days' holiday	100	0.435	43.5
Reinstatement of staff	100	0.217	21.7
		Value of deal	65.2

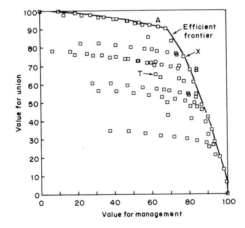

Figure 12.8 Identifying the efficient frontier for the management–union negotiations

position of the other party. This means that if an agreement does not lie on the efficient frontier then other agreements are available which would either offer an improved position to both parties or allow one party to improve its position without doing so at the expense of the other.

It can be seen that the tentative deal (indicated by T on the graph) did not lie on the efficient frontier. All the agreements on the frontier between A and B were more efficient, and it was therefore in the interests of both parties to try to improve on T by reaching one of these agreements. After much more bargaining they eventually settled on deal X. This involved a 7% pay rise but only one day's extra holiday per year, though the staff who had been fired would still be reinstated. Since the deal had a value of 77.4 for the management and 74.9 for the union it represented a clear improvement over the tentative deal for both parties.

It is obviously tempting to ask whether decision analysis can be used to determine an 'optimum' agreement from those available on the efficient frontier. The problem is that in identifying such an agreement we would have to compare the preference scales of the two parties. As we saw in the previous chapter when we considered interpersonal comparisons of preferences, such a comparison would require us to ask whether a swing from a value of 0 to 100 on the management's scale was greater or smaller than a similar swing on the union scale. For example, it may be that the management are far less concerned about a swing between these two values than the union, in which case perhaps an efficient deal should be selected which favours the union, but how can such comparisons be made?

In this problem we have assumed that management could make an assessment of the union's preferences and the trade-offs they were prepared to make. Of course, in the atmosphere of intense negotiations with each party using bluff, pretence and what Raiffa[4] calls 'strategic misrepresentation' (i.e. 'exaggerating the importance of what one is giving up and minimizing the importance of what one gets in return') such an assessment may be difficult to make. Nevertheless, in any negotiations it is necessary to form some idea of the other party's position, and analysis may lead to a sharper perception of this. In some situations the parties in a dispute may be prepared to reveal their preferences to a third party, who will then guide them to (or, in some circumstances, impose upon them) an efficient deal.

PRACTICAL APPLICATIONS

Much of the early work in applying multi-attribute value analysis to negotiating problems was carried out by the consultants Decisions and Designs Incorporated (DDI). This company helped US negotiators to formulate a negotiating strategy for the Panama Canal negotiations in 1974. Other applications can be found in Barclay and Peterson.[5]

The analytic approach was also adopted by a member of the US team in the 1978 negotiations between the United States and the Philippines over the status of the US bases on the islands. Raiffa[4] reports the experiences of the team member who argued that formal analysis led to a creative attitude to the negotiations. Rather than focusing purely on their own position and how it could be defended, the team were encouraged to look for ways of obtaining a better deal by trading off interests where their gain was not necessarily the other party's loss.

Phillips[1] reports the experiences of Cameron Peterson, a consultant with DDI. Peterson found that there were a number of advantages to be gained by bringing the decision analysis approach to bargaining problems. Negotiators could prepare in advance and anticipate the positions of other parties and, by developing a clear understanding of the problem, they were able to be flexible and creative during the negotiations. There was also better communication within the negotiating team and between the team and their organization. However, the approach was found to be least effective where negotiators sought to preserve an air of mystery about their bargaining methods and skills.

SUMMARY

In this chapter we first considered the application of decision analysis to problems where a group meets to decide how resources should be allocated between alternative uses. The problem was characterized by a small number of objectives but a very large number of possible courses of action, and a computer was therefore required to help in the comparison of the alternatives. The use of decision analysis facilitated group participation in the decision process so that the conflict between centralization and local decision making could be resolved. We then showed how decision analysis can help decision makers who are involved in negotiations to identify improved deals which are in the interests of both parties.

DISCUSSION QUESTIONS

(1) What is the commons dilemma and how can decision analysis help to resolve it?
(2) Explain the distinction between within-criterion and across-criteria weights in resource allocation models.
(3) A group of executives have to allocate investment funds between six possible product lines. Their objectives are: (i) to maximize the *growth* of the company, (ii) to maximize *export* earnings, (iii) to maximize *profit* and (iv) to minimize *risk*.

After much debate the group decide to assign the following across-criteria weights: Growth, 100; Exports, 10; Profit, 80; Risk, 20. Discuss how these weights could have been derived and explain what they mean.
(4) Will the use of a decision analysis resource allocation model in a decision conference necessarily lead to an optimum allocation of resources?
(5) Explain why sensitivity analysis is important when using resource allocation models in a decision conference.

(6) What are the potential benefits of using decision analysis in negotiation problems?

(7) What is meant by an efficient deal in the context of negotiation problems?

(8) In what circumstances will the additive value model be inappropriate when modeling a negotiator's preferences?

REFERENCES

1. Phillips, L. D. (1989) Decision Analysis in the 1990's, in A. Shahini and R. Stainton (eds) *Tutorial Papers in Operational Research 1989*, Operational Research Society.

2. Phillips, L. D. (1989) People-centred Group Decision Support, in G. Doukidis, F. Land and G. Miller (eds) *Knowledge-based Management Support Systems*, Ellis Horwood, Chichester.

3. Barclay, S. (1988) *A User's Manual to EQUITY*, Decision Analysis Unit, London School of Economics, London.

4. Raiffa, H. (1982) *The Art and Science of Negotiation*, Harvard University Press, Cambridge, Mass.

5. Barclay, S. and Peterson, C. R. (1976) *Multiattribute Utility Models for Negotiations*, Technical Report DT/76-1, Decisions and Designs Inc., McLean, Virginia.

Chapter 13

Alternative Decision-support Systems

INTRODUCTION

Up to now we have focused on a single decision-aiding technique, decision analysis. In this chapter we present an overview of two alternative ways of aiding decision making: linear modeling and expert systems. *Linear modeling* involves building a statistical model of a person's judgments or predictions and subsequently utilizing the model instead of the person. *Expert systems* relate to building a model of the decision processes of an expert decision maker. In a similar way to linear modeling, the expert system representation of the decision maker is subsequently used instead of the person.

From the above short overviews it is clear that the three decision-aiding technologies—decision analysis, linear modeling and expert systems—require the elicitation and representation of human judgment. As we shall see, linear modeling and expert systems place different emphasis on the assumed quality of the judgmental input. In this chapter we will first introduce the decision-aiding technologies and then compare and contrast them, both with each other and with decision analysis. Our focus will be on the domains of applicability of the different approaches and on the validity of the resulting decisions.

EXPERT SYSTEMS

What is an Expert System?

Expert systems are one offshoot of research into artificial intelligence (AI). The aim of AI is to represent the totality of human intelligence and thought within a computer system. Within AI research are such fields of study as:

1. *Voice/image recognition.* The aim here is to produce systems that can recognize verbal instructions and visual images, such as the changing view from the driving seat of a car. As yet, systems can only recognize a few score of spoken words which are given as one-word commands by an individual whose voice patterns have been matched by the computer. In addition, computer systems have been developed that can recognize simple images such as typewriter typeface print, but a system that can perform the same voice- and image-recognition functions as a copy typist may be an unrealizable dream. Consider, for example, the problems involved in producing a typewritten letter from verbal/written instructions for a person the typist has never met before—the recognition of an unfamiliar voice and/or handwriting and of the person's face when he or she calls to collect the completed work would confound any computer system currently envisaged.

2. *Robotics.* While applications are now increasingly common in manufacturing for example, the ultimate objective of robotics is to produce machines that 'think' and 'act' like humans. Recall, for instance, the intelligent thoughts and actions of C3PO and the other *Star Wars* robots. If producing a copy-typing system such as the one mentioned above is known to present unsurmountable difficulties to any conceivable computer system based on known technology, then robotic research to produce machines with the flexibility of people presents even more complex problems to the AI researcher.

Early research on expert systems was also focused on relatively complex problems such as diagnosing the disease from which a person is suffering. The aim was for the system to perform the diagnosis in the same way as an expert physician. However, diagnosis turned out to be a difficult problem, and even now, after many person-years of effort, none of the systems built are in routine use. One reason for this is that the systems were developed by academics who were more interested in producing academic papers to further their careers. Relatively simple practical problems that can be solved easily hold no challenges (or publications) and so tend to be avoided by university-based researchers. More recently, *commercial* advantage has been seen in picking the 'low-hanging fruit', and it is these expert systems, built in person-months rather than person-years, that form the focus of this section. As we shall see, they are often targeted on a particular area of expertise.

Several definitions of an expert system have been proffered, but the Expert Systems group of the British Computer Society provides a generally agreed and workable definition that it *is the modeling, within a computer, of expert knowledge in a given domain, such that the resulting system can offer intelligent advice or take intelligent decisions.* One important addition to this definition is that the system should be able to *justify* the logic and reasoning underlying its advice or decision making.

It follows that expert systems act as *decision aids* or *decision-support systems* when operating by giving *advice* to a (non-expert) human decision maker. Expert systems can also act as *decision makers* without any human–computer dialogue or interaction. Finally, expert systems can act as *trainers* by instructing human novices to become experts in a particular area of expertise.

Three brief sketches will give the essence of these distinctions:

1. A fisherman goes into a tackle shop. He wants to catch a certain type of fish in the particular river conditions where he fishes. The expert system questions him about the fish, river conditions and weather, etc. and *advises* him to use a certain type of fly or bait.
2. In a factory, the sensors within a machine indicate an imminent component failure. The expert system *decides* to close the machine down and alert a particular fitter to attend the problem. At the same time, the expert system orders the required spare part from the store room and dispatches it to the fitter.
3. A school student works through mathematical problems with an expert system. After the student has worked through a series of problems the system diagnoses the underlying cause of the observed errors, gives the student some extra problems to confirm the diagnosis and then proceeds to give *tuition* in the required mathematical skills.

What is Expert Knowledge?

The nature of human knowledge is an area of much debate and controversy. However, it assumes a more concrete form in the practice of *knowledge engineering*. For expert systems, this is the skill of obtaining and manipulating human knowledge so that it can be built into a computer model which in some ways behaves like an expert. Gaining the knowledge from the expert, an initial focus of knowledge engineering, is termed *knowledge elicitation*, and this is usually concerned with obtaining knowledge from *people* rather than documents. In fact, the knowledge of experts goes far beyond that contained in textbooks. For example, Wilkins *et al.*[1] cite the case of medical expertise where, despite years of textbook study, students are unable to show *diagnostic expertise*. This is achieved from an 'apprenticeship period' where they *observe* experts in real diagnoses and attempt to duplicate the skill by practicing themselves. Indeed, expert knowledge consists of many *unwritten* 'rules of thumb':

> [it is] . . . largely heuristic knowledge, experimental, uncertain—mostly 'good guesses' and 'good practice', in lieu of facts and figures. Experience has also taught us that much of this knowledge is private to the expert, not because he is unwilling to share publicly how he performs, but because he is unable to.
>
> He knows more than he is aware of knowing . . . What masters really know is not written in the textbooks of the masters. But we have learned that this private knowledge can be uncovered by the careful, painstaking analysis of a second party, or sometimes by the expert himself, operating in the context of a large number of highly specific performance problems.

Sometimes in order to understand one expert's actions the expertise of another is required. For example, in organized human/machine chess matches a high-ranking player is often present in order to explain the likely reason for each player's moves. Similarly, in eliciting medical expertise a doctor can be employed to observe a patient–doctor interview and infer the reasons for questions asked of the patient.

Given the 'hidden' nature of expert knowledge, it is not surprising to find research in the area of knowledge engineering pointing to the difficulties of elicitation. Hayes-Roth et al.[2] have described it as a 'bottleneck in the construction of expert systems'. For example, communication problems arise because not only is the knowledge engineer relatively unfamiliar with the expert's area or 'domain' but the expert's vocabulary is often inadequate for transferring expertise into a program. The 'engineer' thus plays an intermediary role with the expert in extending and refining terms. Similarly, Duda and Shortcliffe[3] conclude that:

> The identification and encoding of knowledge is one of the most complex and arduous tasks encountered in the construction of an expert system . . . Thus the process of building a knowledge base has usually required an AI researcher. While an experienced team can put together a small prototype in one or two man-months, the effort required to produce a system that is ready for serious evaluation (well before contemplation of actual use) is more often measured in man-years.

Wilkins et al.[1] reinforce this view and note that attempts to automate the 'tedious' and 'time consuming' process of knowledge acquisition between expert and 'engineer' have so far proved unsuccessful. It is clear that knowledge elicitation for expert system development shares many characteristics of knowledge elicitation for decision analysis, discussed in Chapter 5.

How is Expert Knowledge Represented in Expert Systems?

Having completed the difficult process of elicitation, the knowledge must be represented in a form that can be implemented in a computer language.

This is most commonly achieved in the form of *production rules*. For example:

IF a car is a VW beetle THEN the car has no water-cooling system.

More formally:

IF (condition in database) THEN (action to update the database)

Production rules can have multiple conditions and multiple actions. The action of a production rule may be required to ask a question of the user of the system or interact with a physical device in addition to updating the database. Production rule-based expert systems often use many hundreds of rules and so control of their action becomes a serious problem for the knowledge engineer.

The *control structure* determines what rule is to be tried next. The control structure is often called the rule interpreter or *inference engine*. In response to information gained from the user in interaction with the expert system, the inference engine selects and tests individual rules in the rule base in its search for an appropriate decision or advice. It usually does this by *forward chaining*, which means following pathways through from known facts to resulting conclusions. *Backward chaining* involves choosing hypothetical conclusions and testing to see if the necessary rules underlying the conclusions hold true. As an added complication, we note that the rules elicited from experts often contain a degree of uncertainty. For example:

IF a car won't start THEN the cause is *likely* to be a flat battery but it *could be* lack of fuel and *might be* . . .

Most expert systems that can tolerate uncertainty employ some kind of probability—like a measure to weigh and balance conflicting evidence. Before providing an example of an expert system application it is important to recognize the significance of the user–system interface in systems design. Expert systems are often used by non-experts, many of whom may also be unfamiliar with computers. Successful systems must be able to interface effectively with their users in order to:

1. Gain the information needed to test the rules;
2. Give understandable advice in plain English and justify the logic and reasoning underpinning the advice proffered or decision made.

Psychologists, rather than computer programmers, have the sort of skills necessary to build appropriate interfaces. Consequently, successful knowledge engineers integrate both computing and psychological skills.

Overall, expert systems are often developed to reproduce experts' decision-making processes in relatively narrow speciality areas. The way in which expertise is represented in the systems has most often been in terms of production rules, since these are easily programmable. Two types of expert systems can be distinguished. The first are basically academic research projects where difficult, or potentially unsolvable, problems are tackled so that new ways of representing or eliciting knowledge must be developed. The second set of systems are those built by consultants utilizing commercially developed *expert system shells*. These are easily programmable in the same way that word-processing programs and spreadsheets provide easy-to-use tools for the office environment. Consultant-built expert systems have tended to focus on problems where uncertainty is not present. This is because Bayes' theorem is not easily understood by non-statisticians. Even for statisticians, the computations become complex when data contain dependencies. Indeed, in practical applications of expert system technologies, expert judgment is often represented in terms of decision trees without uncertainty nodes. Such tree representations of knowledge lend themselves to straightforward programming.

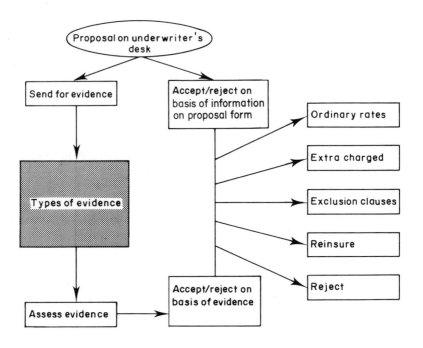

Figure 13.1 Underwriting options

An Example of an Expert System Application in Life Underwriting

Bolger et al.[4] have built an expert life-underwriting system that was the result of a collaborative project between the Bristol Business School and the Clerical Medical Investment Group, a major UK life assurer. The system's major purpose was to enable novice underwriters to underwrite proposals to the same level of accuracy as senior underwriters.

In order to do this, they had to model the decision-making procedures used by a senior underwriter in assessing the information contained in a life-proposal form. Figure 13.1 gives the underwriting options.

Bolger et al. built the system in six modules: occupation, geography, lifestyle/AIDS, financial, hobbies and medical. Each module contained the rules that the senior underwriter used to assess risk. Figure 13.2 presents a small example of the rule base of the geographical module. Note the similarity between decision tree representations in decision analysis and the representation of the sequence of rule testing in the expert system.

The knowledge elicitation was performed between two knowledge engineers and one 'expert'. Knowledge elicitation techniques included interviews, *card sorting* and *context focusing*.[5] Card sorting consisted of the knowledge engineer writing down on cards the names of, say, countries. In one version of the card-sorting technique the expert chose three countries at random (the cards were face down) and then had to sort them into two groups so that the countries named on two of the cards were more similar to each other in some respect than to the third country. In this way the knowledge engineer was able to explore the way in which an underwriter views countries in terms of risk dimensions. Context focusing consisted of the knowledge engineer role playing a novice underwriter who had in front of him a completed life-proposal form. The senior underwriters' task was to help the novice come to an underwriting decision by means of telephone communication. The sequence of rule testing engaged by the expert was recorded by the knowledge engineer and provided one means of identifying the priority of rules. One set of rule testing with high priority is given below:

Has every question on the proposal form been answered?
IF yes THEN ask:
Is the current proposal sum insured within no evidence limits?
IF yes THEN ask:
Are there any previous sums assured in force?
IF no THEN ask:
Are height and weight within acceptable parameters?
IF yes THEN ask:
Are there any other questions on the proposal form that are answered 'yes'?
IF no THEN ACCEPT PROPOSAL

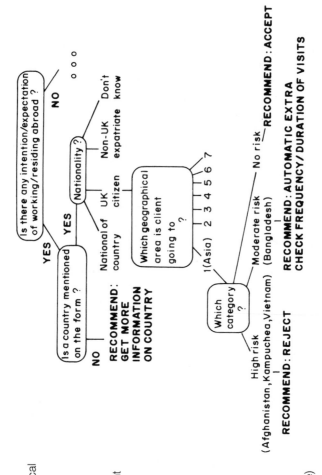

If applicant intends/expects to work/reside abroad
THEN check if his or her destination is indicated on the proposal form
ELSE check for some other geographical details . . . .

IF country applicant intends to visit is indicated on proposal form
THEN look for indicators of the applicant's nationality on the form
ELSE get information from the applicant about the country s/he is destined for (exit module)

Following branch for UK citizen destined for Asia only:

IF country indicated is Afghanistan, Kampuchea, or Vietnam ('high risk')
THEN recommendation is to reject the proposal (exit module)

IF country indicated is Bangladesh ('moderate risk')
THEN recommendation is to charge an extra depending on the frequency/duration of visits (exit module)
ELSE any other Asian country recommendation is to accept the proposal with no geographical extras (exit module)

Figure 13.2

Overall, it is clear that many of the functions of organizations involve expertise. In addition, some individuals' expertise is valued more than others and expert systems have the potential to make the most valued expertise within a company available immediately on demand. Benefits to be gained include:

1. Improved company image through more efficient service;
2. Improvement in the quality and consistency of decision making;
3. Release of experts in a particular field to concentrate on more complex or critical problems;
4. Better communication of knowledge across the organization;
5. Provision of an accessible reference source for crucial knowledge;
6. Improving the quality and cost effectiveness of training.

In contrast to the approach of expert systems, the work on statistical models of judgment has a basic premise that people cannot give accurate descriptions of how they arrive at judgments or decisions. The methodology is, simply, to model via linear regression, the relationship between what judgments a person makes and the information upon which the judgments are based. We turn to this work next.

STATISTICAL MODELS OF JUDGMENT

One of the major databases used for early experimentation on multi-attributed inference has been that collected by Meehl.[6] The judgmental problem used was that of differentiating psychotic from neurotic patients on the basis of their MMPI questionnaire profiles.

Each patient, upon being admitted to hospital, had taken the MMPI. Expert clinical psychologists believe (or at least used to believe) that they can differentiate between psychotics and neurotics on the basis of profile of the eleven scores. Meehl noted that

> because the differences between psychotic and neurotic profile are considered in MMPI lore to be highly configural in character, an atomistic treatment by combining scales linearly should be theoretically a very poor substitute for the configural approach.

Initially, researchers tried to 'capture' or 'model' expert judges by a simple linear regression equation. These judgmental representations are constructed in the following way. The clinician is asked to make his diagnostic or prognostic judgment from a previously quantified set of cues for each of a large number of patients. These judgments are then used as the dependent

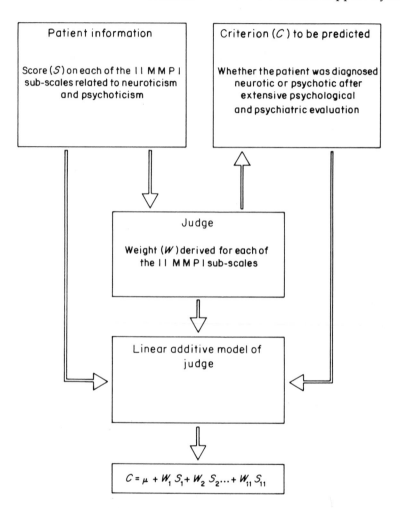

Figure 13.3 Basic paradigm for the construction of a linear additive model of a judge

variable in a standard linear regression analysis. The independent variables in this analysis are the values of the cues. The results of such an analysis are a set of regression weights, one for each cue, and these sets of regression weights are referred to as the judge's 'model' or 'policy'. Figure 13.3 sets out the basic paradigm for a study of multi-attributed inference.

How do these models make out as predictors themselves? That is, if the regression weights (generated from an analysis of one clinical judge) were used to obtain a 'predicted score' for each patient, would these scores be more valid, or less valid, than the original clinical judgments from which the regression weights were derived? To the extent that the model fails to capture valid non-linear variance to the judges' decision processes, it should perform worse than the judge; to the extent that it eliminates the random error component in human judgments, it should perform better than the judge.

What were the results of this research? The overwhelming conclusion was that the linear model of the judge's behavior outperformed the judge. Dawes[7] noted:

> I know of no studies in which human judges have been able to improve upon optimal statistical prediction . . . A mathematical model by its very nature is an abstraction of the process it models; hence if the decision-maker's behavior involves following valid principles but following them poorly these valid principles will be abstracted by the model.

Goldberg[8] reported an intensive study of clinical judgment, pitting experienced and inexperienced clinicians against linear models and a variety of non-linear or configural models in the psychotic/neurotic prediction task. He was led to conclude that Meehl chose the wrong task for testing the clinicians' purported ability to utilize complex configural relationships. The clinicians achieved a 62% rate, while the simple linear composite achieved 70%. A 50% hit rate could have been achieved by chance as the criterion base rate was approximately 50% neurotic, 50% psychotic.

Dawes and Corrigan[9] have called the replacement of the decision maker by his model 'bootstrapping'. Belief in the efficacy of bootstrapping is based on a comparison of the validity of the linear model of the judge with the validity of his or her holistic judgments. However, as Dawes and Corrigan point out, that is only one of two logically possible comparisons. The other is between the validity of the linear model or the judge and the validity of linear models in general. That is, to demonstrate that bootstrapping works because the linear model catches the essence of a judge's expertise and at the same time eliminates unreliability, it is necessary to demonstrate that the weights obtained from an analysis of the judge's behavior are superior to those that might be obtained in another way—for example, obtained randomly.

Dawes and Corrigan constructed semi-random linear models to predict the criterion. The sign of each predictor variable was determined on an *a priori* basis so that it would have a positive relationship to the criterion.

On average, correlations between the criterion and the output predicted from the random models were higher than those obtained from the judge's models. Dawes and Corrigan also investigated equal weighting and discovered that such weighting was even better than the models of the judges or the random linear models. In all cases, equal weighting was superior to the models based on judges' behavior.

Dawes and Corrigan concluded that the human decision maker need specify with very little precision the weightings to be used in the decision— at least in the context studied. What must be specified is the variables to be utilized in the linear additive model. It is precisely this knowledge of 'what to look for' in reaching a decision that is the province of the expert clinician.

The distinction between knowing what to look for and the ability to integrate information is illustrated in a study by Einhorn.[10] Expert doctors coded biopsies of patients with Hodgkin's disease and then made an overall rating of severity. These overall ratings were very poor predictors of survival time, but the variables the doctors coded made excellent predictions when utilized in a linear additive model.

These early studies cast doubt on judgmental forecasting abilities. However, more recent studies present contradictory evidence, and it is to these that we turn next.

Recent Research

Many recent studies, conducted in real-world settings, have looked at bankruptcy prediction. Libby,[11] in a major study, had 43 experienced loan officers make predictions for 60 real but disguised companies, half of which had failed. These predictions were made on the basis of the limited financial information contained in five financial ratios. Other information such as absolute amount of income, notes to the accounts, etc. was excluded from the experimental study. Nevertheless, the mean predictive accuracy of the loan officers' judgments was high, at 74%. However, in this artificially limited study only nine of the 43 judges did better than the ratio of assets to liabilities (see Dawes[12] for an insightful discussion of the issues).

Whitred and Zimmer[13] point out that, in principle, loan officers may outperform a linear model by the valid use of non-linear relationships between ratios and (non) bankruptcy. However, the robustness of the models to violations of non-linearity will make this potential advantage of man over model practically immaterial. For loan officers to systematically outperform the model they must have access to information unavailable to the model, information which may be prevalent in real life rather than in laboratory situations.

In fact, Shepanski[14] reported an experiment to test a linear representation and various non-linear representations of information-processing behavior in the task of credit evaluations. Participants in the experiment were presented with sets of information describing prospective business borrowers in terms of payment record, financial condition and quality of the company's management. Shepanski argued that the credit judgment task is best represented by a non-linear model. Additionally, in real-life credit valuations the composition and size of the information employed will change. Information gathering is costly and, for example, applications for a large loan will entail a much more comprehensive credit investigation than a small loan application. Such flexibility in information search cannot be captured by statistical modeling that is better suited to repetitive forecasts with a static number of predictor variables. However, as Dawes *et al.*[15] have pointed out, the small number of studies that have provided clinicians with access to preferred sources of information have generally shown the superiority of the statistical model. As these authors note, human judgment can theoretically improve on statistical modeling by recognizing events that are not included in the model's formula and that countervail the actuarial conclusion. Dawes *et al.* argue that such events are rare but, as we have already shown in Chapter 9, this is the exact situation where forecasting practitioners advocate the need for judgment. Indeed, recent studies[16,17] provide evidence of the quality of human judgment compared to statistical model when 'broken leg' cues are part of the information available for decision making. The term 'broken leg cue' is due to Meehl.[6] He noted that the knowledge that a certain person had just broken his or her leg would invalidate any predictions of a person's movements (e.g. to the theater, particular restaurants, etc.) based on historic statistical data.

To illustrate, Chalos[18] investigated the ability of an outcome-based credit-scoring model to assess financial distress and compared the performance of the model with that of loan review committees and individual loan officers. The major finding was the loan review committees significantly outperformed the model and the individual officers. The model was a stepwise discriminant model built using eight financial ratios as cue variables. The loan review officers/committees had additional information for each judgment in the previous three years' financial statements. Chalos's results indicated that loan committees may be beneficial, and the additional time required may be more than offset by the reduction in loan cost errors. In a related study, Casey and Selling[19] used MBA students as subjects in a bankruptcy prediction task and noted that if a firm's specific financial data do not provide a clear-cut signal of its financial viability, then subjects would be expected to incorporate available prior probability information into their judgment processes. Such additional information is, of course, likely to be available in the everyday situations of loan officers.

Although early studies of linear modeling in clinical settings showed evidence that the model of the judge outperforms the judge on whom the model was based, more recent evidence is contradictory. This shows that experts can recognize the significance of extra-model information and use it appropriately. Such characteristics of experts can, potentially, be captured in expert system representations of knowledge.

COMPARISONS

Essentially, *decision analysis* is utilized for unique or one-off decisions under uncertainty. The decision maker provides the decision analyst with the temporal sequencing of possible acts and events (the decision tree), his or her opinion about the likelihood of events (subjective probabilities) and his or her subjective valuation (utilities) of the consequences of particular act and event combinations or outcomes. The whole approach is predicated on the notion that decomposition and subsequent recomposition of a decision problem, using decision trees, will improve decision making. As we have seen, the implicit theory is that we humans have limited information-processing capacity and that the expected utility computations are best left to the analyst's computer. Nevertheless, decision analysis still makes the assumption that the decision maker's prime inputs of subjective probability and utility have validity. Recall that the practice of sensitivity analysis focuses elicitation methodologies on 'critical' assessments. In a similar manner, the bootstrapping approach involves the assumption that decision makers are able to identify the key predictor variables to be entered into the prediction equation. Optimal weighting of the predictor variables' impact on the prediction equation is best left to the statistical modeling techniques. In contrast to decision analysis, bootstrapping models are best deployed in repetitive decision-making situations where only scores on the predictor variables vary from one prediction to another.

In more dynamic environments, where fresh predictor variables may be expected to be added to the cue variable set or where the possibility of 'broken leg' cues occurring is pronounced, bootstrapping systems will be less successful and may be best overridden by holistic judgment. This conclusion complements that of Chapter 9, where we argued for the validity of judgmental adjustments to the output of extrapolative and econometric statistical forecasting models.

Expert systems, in contrast, are predicated on the assumption that expert, informed *holistic* decision making is valid[20]. Conventional approaches to assessing the adequacy of a system focus on the convergence between the system's decision/diagnosis/advice and that of the expert who is modeled in the system. Although expert decision making is conventionally decomposed into if/then production rules, no normative theory or statistical technique oversees the aggregation or selection of these rules into an optimal set or sequence for execution when the system is used.

In common with bootstrapping models, expert systems are most useful in repetitive decision or advice-giving situations. The reason for this is simple: if the conditions that lead to a decision or piece of advice are rarely encountered then the knowledge engineering time needed to model that 'leg' of the decision tree may not be cost effective. In many commercial applications it is far better if the need for extra-system human expertise is recognized by the system and a complex problem is handed over to an expert for resolution. For example, in the life underwriting system outlined above, infrequent combinations of medical conditions and medical treatments that *could*, potentially, indicate that a proposal concerns a poor life risk are dealt with by the chief underwriter.

Overall, there are several differences in the domain of applicability of decision analysis, bootstrapping models and expert systems. One commonality to all is the primacy of human judgment. As we saw in Chapter 9, human judgment is likely to be good when practice and useful feedback provide conditions for the quality of judgment to be evaluated. In decision analysis practice the decision analyst working on (what is usually) a unique decision can only check the *reliability* of the decision maker's inputs of probability and utility. Questions to do with the validity (e.g. calibration) of the assessments are much more difficult to evaluate for one-off assessments given by non-practiced assessors. Fortunately, sensitivity analysis provides a fallback that at least allows identification of critical inputs. Decision conferencing techniques allow further analysis and discussion of these inputs. Clearly, in the absence of 'the truth' an achievable alternative of a group consensus or, at least, knowledge of the variability in the groups' estimates is useful knowledge.

We have argued previously that decisions, once made, are often 'made' to work. For this reason, questions to do with the validity of decision analyses are often raised but seldom answered. Most often, the question of validity is sidestepped and questions concerning the 'valuation' of decision analysis are raised instead, as we saw in Chapter 11 when considering decision conferencing. Questions on the validity of linear modeling are more easily answered, since the method is most useful under conditions of repetitive decision making. As we have seen, this method has shown evidence of incremental validity over the holistic judgments/predictions of the judge on whom the model was based.

Questions relating to the validity of expert systems have often not been asked. Most often expert system researchers have been concerned with problems of redundancy, conflicts, circularity (e.g. self-referencing chains of inference) and incompleteness in rule sets.[21] When the validity issue is analyzed the level of analysis is usually a comparison of the expert's and the completed expert system's decisions or advice when both are presented with example cases. The resulting error rate, i.e. the proportion of 'incorrect' decisions or advice given by the system, is often summarized by a simple percentage count. However, as we have seen in our discussion of linear models, human judgment contains a random error component. The benefit of linear modeling is elimination of this error by averaging techniques. Hence the incremental validity of the model of the judge over the judge's holistic decisions/predictions. Methods of validating expert systems have not, to date, been able to systematically identify and extract the random error component in human judgment. The only method at the knowledge engineer's disposal with which to identify such a component within the expert system representation of the expert's knowledge is to ask the expert to introspect on the rule set.

SUMMARY

In this chapter we introduced two additional ways of aiding decision making: expert systems and linear models. In common with decision analysis, these decision-aiding technologies involve a substantial component of judgmental modeling but are applied in different circumstances and place differing emphases on the nature and assumed validity of the judgmental components.

REFERENCES

1. Wilkins, D. C. *et al.* (1984) Inferring an Expert's Reasoning by Watching, *Proceedings of the 1984 Conference on Intelligent Systems and Machines.*
2. Hayes-Roth, R. *et al.* (1983) *Building Expert Systems*, Addison-Wesley, Reading, Mass.
3. Duda, R. O. and Shortcliffe, E. H. (1983) Expert Systems Research, *Science,* **220**, 261–268.
4. Bolger, F., Wright, G., Rowe, G., Gammack, J. and Wood, R. J. (1989) LUST for Life: Developing Expert Systems for Life Assurance Underwriting, in N. Shadbolt (ed.) *Research and Development in Expert Systems VI*, Cambridge University Press, Cambridge.
5. For details see Wright, G. and Ayton, P. (1987) Eliciting and Modelling Expert Knowledge, *Decision Support Systems*, **3**, 13–26.
6. Meehl, P. E. (1957) When Shall We Use our Heads instead of the Formula? *Journal of Counselling Psychology*, **4**, 268–273.
7. Dawes, R. M. (1975) Graduate Admission Variables and Future Success, *Science*, **187**, 721–743.
8. Goldberg, L. R. (1965) Diagnosticians versus Diagnostic Signs: The Diagnosis of Psychosis versus Neurosis from the MMPI, *Psychological Monographs*, **79**, 602–643.
9. Dawes, R. M. and Corrigan, B. (1974) Linear Models in Decision-making, *Psychological Bulletin*, **81**, 95–106.
10. Einhorn, H. J. (1972) Expert Measurement and Mechanical Combination, *Organizational Behavior and Human Performance*, **7**, 86–106.
11. Libby, R., Accounting Ratios and the Prediction of Failure: Some Behavioral Evidence, *Journal of Accounting Research*, Spring, 150–161.
12. Dawes, R. M. (1979) The Robust Beauty of Improper Linear Models, *American Psychologist*, **34**, 571–582.

13. Whitred, G. and Zimmer, I. (1985) The Implications of Distress Prediction Models for Corporate Lending, *Accounting and Finance*, **25**, 1–13.
14. Shepanski, A. (1983) Tests of Theories of Information Processing Behavior in Credit Judgement, *The Accounting Review*, **58**, 581–599.
15. Dawes, R. M., Faust, D. and Meehl, P. (1989) Clinical versus Actuarial Judgement, *Science*, **243**, 1668–1673.
16. Johnson, E. J. (1988) Expertise and Decision under Uncertainty: Performance and Process, in M. T. H. Chi, R. Glaser and M. J. Farr (eds) *The Nature of Expertise*, Erlbaum, Hillsdale, NJ.
17. Blattberg, R. C. and Hoch, S. J. (1989) Database Models and Managerial Institution: 50% Model and 50% Manager, Report from the Center for Decision Research, Graduate School of Business, University of Chicago, May.
18. Chalos, P. (1985) The Superior Performance of Loan Review Committee, *Journal of Commercial Bank Lending*, **68**, 60–66.
19. Casey, C. and Selling, T. I. (1986) The Effect of Task Predictability and Prior Probability Disclosure on Judgment Quality and Confidence, *The Accounting Review*, **61**, 302–317.
20. For a recent review see Wright, G. and Bolger, F. (1991) *Expertise and Decision Support*, Plenum, New York.
21. Nazareth, D. L. (1989) Issues in the Verification of Knowledge in Rule-based Systems, *International Journal of Man–Machine Studies*, **30**, 255–271.

Suggested Answers to Selected Questions

CHAPTER 2

(2) (a) Ultraword, Easywrite and Super Quill are on the efficient frontier.
(b) Easywrite has the highest aggregate value of 77.25 (this value is obtained after normalizing the weights).
(c) This implies that the two attributes are not mutually preferential independent, so the additive model may not reflect your preferences accurately.

(3) (a) Design A will offer the highest aggregate value for benefits as long as the weight for environmental impact is below about 11. If the weight is higher than this then Design C offers the highest valued benefits.
(b) Designs A, C and D are on the efficient frontier.
(c) The manager is prepared to pay $4000 for each extra benefit point (i.e. $120 000/30). A switch from Design D to A would cost only $731.7 for each extra benefit point (i.e. $30 000/41) and would therefore be worth making. However, a switch from A to C would cost $4705.8 per extra benefit point (i.e. $80 000/17) and would therefore not be worth making. Therefore choose Design A.

(4) (a) Rail/ferry has the highest value for aggregate benefits, i.e. 81.
(b) Rail/ferry and road/ferry lie on the efficient frontier.
(c) The manager is prepared to pay $1167 for each extra benefit point (i.e. $70 000/60). A switch from road/ferry to rail/ferry would cost $567 for each extra benefit point (i.e. $30 000/53) and is therefore worth making. Therefore choose rail/ferry.

(5) (c) Values: Inston, 56; Jones Wood, 66; Peterton, 36.8; Red Beach, 46.4; Treehome Valley, 43.6.
(d) Jones Wood and Red Beach lie on the efficient frontier.
(e) Jones Wood has the highest aggregate benefits whatever weight is assigned to visual impact.

CHAPTER 3

(1) (a) Assuming that the classical approach is valid: 120/350.
(b) Assuming that the relative frequency approach is valid: 8/400.
(c) 0.5 using the classical approach, though the relative frequency approach suggests about 0.515 in some Western industrialized countries.
(d) Assuming that the relative frequency approach is valid: 21/60.
(e) This will be a subjective probability.
(2) (a) 0.25; (b) 0.6; (c) 0.95.
(3) (a) 64/120; (b) 79/120; (c) 67/120; (d) 85/120; (e) 74/120.
(4) (a)(i) 41/120; (ii) 18/64; (iii) 23/56; (iv) 53/120; (v) 32/64.
(5) (a)(i) 40/100; (ii) 30/100; (iii) 45/100; (iv) 25/30; (v) 25/40.
(6) (a) 0.001; (b) $0.9 \times 0.95 \times 0.8 = 0.684$
(7) (a) 0.192.
(8) 0.48.
(9) 0.00008.
(10) 0.6.
(11) (a) 0.54; (b) p(Kingstones only) + p(Eadleton only) = $0.06 + 0.12 = 0.18$.
(12) (a) 2.76 requests; (b) discrete.
(13) $94 000.
(14) (a) $0: 0.4; $40: 0.252; $50: 0.126; $60: 0.042; $80: 0.108; $100: 0.054; $120: 0.018.
(b) $35.1.

CHAPTER 4

(1) Option 2 has the highest expected profit of £24 000.
(2) The speculator should purchase the commodity (expected profit = $96 000).
(3) Carry one spare (expected cost = $5400).
(5) (a) Bid $150 000 (expected payment = $90 000); (b) Bid $100 000 (expected utility = 0.705, assuming a 0 to 1 utility scale).
(7) The Zeta machine (expected utility 0.7677).
(8) (b) Choose the metal design (expected utility 0.7908, assuming a 0 to 1 utility scale).

CHAPTER 5

(1) (b) Invest in the development and, if it is successful, go for large-scale production (expected returns = $1.65 million).

(c) Do not invest in the development if the probability of success is less than about 0.387.

(d) Not investing in the development now has the highest expected utility (0.6 as against 0.5625 if the development goes ahead). This is true as long as the probability of a successful development is less than 0.64.

(2) The engineer should attempt to repair the machine himself and, if necessary make a second attempt. (Note, however, that the decision is very close: the expected cost of attempting the repair himself is $30 780, as opposed to $30 800 if the specialist local company is called in immediately. Sensitivity analysis is therefore advisable.)

(3) (b) Westward should *not* bring the launch forward (expected profit = $3.005 million, as opposed to $2.68 million for bringing the launch forward and $0 for not launching at all), and if the rival launches first they should increase their level of advertising.

(c) The policy is totally insensitive to changes in these probabilities, i.e. not bringing the launch forward is the best option whatever estimates are used for the probabilities of beating the rival.

(4) (b) The Authority should erect a cheap temporary barrier, but if the barrier is damaged they should *not* repair it (the expected cost of this policy is $1.275 million as opposed to $1.48 million for doing nothing and $1.716 million for erecting an expensive barrier).

(5) (a) The 20-person team gives the lowest expected costs of $11 600.

(b) The manager should now use a 15-person team and hire the equipment only if the overhaul is behind schedule on the Saturday evening. (Note that the two expected costs are very close, $11 400 for the 15-person team and $11 600 for the 20-person team, which suggests that sensitivity analysis should be carried out.)

CHAPTER 6

(1) (a) Profit probability distribution is: $0: 0.08; $100: 0.20; $200: 0.24; $300: 0.30; $400: 0.18.

(c) Probability distribution estimated from simulation is: $0: 0; $100: 0.20; $200: 0.30; $300: 0.30; $400: 0.20.

(3) (b) Assuming that the mean–standard deviation screening procedure is valid, only designs 1 and 6 lie on the efficient frontier. Design 6 offers higher returns but also has a higher level of risk than design 1.

(5) (a) The option of replacing the plant with new equipment exhibits first-degree stochastic dominance over the option of extending the existing plant.

(b) Replacing the plant with new equipment also exhibits second-degree stochastic dominance over the option of moving the company's operations.

CHAPTER 7

(1) (i) p(high sales)=0.7; p(low sales)=0.3.
   (ii) Posterior probabilities: p(high sales)=0.4375; p(low sales)=0.5625.
(2) p(sales exceed one million units)=0.4615.
(3) p(machine accidentally overfilled)=0.2963.
(4) p(minerals in commercial quantities)=0.8182.
(5) (a)(i) Build new plant (expected NPV=\$450 000); (ii) EVPI=\$85 000.
   (b) The company should now expand existing plant (expected NPV=\$390 750).
(6) (a)(i) Plan for medium sales (expected profit=\$164 000);
   (ii) EVPI=\$64 000.
   (b) The company should still plan for medium sales (expected profit=\$152 190).
(7) Expected value of test=\$399.
(8) (a)(i) The product should be launched (expected NPV=\$18m);
   (ii) EVPI=\$12m.
   (b) EVII=\$5.11m therefore it is worth test marketing the product.

Suppliers of Computer Software

Package	Name and Address of Supplier
ARBORIST	Texas Instruments Incorporated 12337 Technology Boulevard Austin, Texas 78759, USA
EQUITY	Decision Analysis Unit London School of Economics Houghton Street London WC2A 2AE, UK
HIVIEW	Decision Analysis Unit (as above)
PREDICT!	Risk Decisions/Unison Technology Inc. Coraopolis, PA, USA
@RISK	Palisade Corporation 2189 Elmira Road Newfield New York, 14867-9444, USA
SUPERTREE	SDG Decision Systems 2440 Sand Hill Road Menlo Park, California 94025-6900, USA

Index

across criteria weights, 267–9
addition rule of probability, 42–3
additive value model, 21–2, 27–9, 260
Adorno, T. C., 210
aggregation
  of benefits, 21–2, 27–9, 269–70, 276–7
  of group judgments, 242–57
Alberoni, F., 240
Allais's paradox, 83–5
anchoring and adjustment heuristic, 194
Angus-Leppon, P., 219
approval voting, 249
ARBORIST, 107, 302
Arkes, H. R., et al., 153
Armstrong, S., 220, 221
Arrow, K. J., 249
Arrow's impossibility theorem, 249
Ashton, A. H., 245, 246
Ashton, R. H., 245, 246
Assmus, G., 185
attribute
  definition of, 8
  proxy, 8–9
availability heuristic, 194, 235–6
axioms
  of probability theory, 54
  of SMART, 26–7
  of utility, 76–9
Ayton, P., 240

Balthasar, H. U., 222
bankruptcy prediction, 292
Barclay, S., 11, 31, 260, 278
Bayes' theorem, 165–75, 198, 247
Beach, L. R., 196, 197, 215, 240
Bell, D. E., 111
Bernoulli, D., 64
biased tests, 166
bisection method, 17
Blattberg, R. C., 297

Bodily, S. E., 29, 109, 110
Bolger, F., 287
bootstrapping, 289, 294
Bordley, R. F., 247
Brigham, E. F., 146
broken leg cues, 293
Brownlow, S. A., 13
Budescu, D. V., 230
Buede, D. M., 10, 11, 94
Bunn, D., 94, 223, 230, 239

calibration, 202, 213, 233, 235
Carbone, R., 220
Casey, C., 293
Casey, C. J., 204
certainty axiom, 54
certainty equivalence method, 79–81
Cerullo, M. J., 223
Chalos, P., 293
chance node, 66, 103
Chapman, J. P., 153
Chapman, L. J., 153
Choisser, R. W., 11
Christensen-Szlanski, J., 196, 197
citation bias, 196
classical approach to probability, 39
clinical judgment, 289
cognitive style, 203
  decisive style, 203
  flexible style, 204
  hierarchic style, 204
  integrative style, 204
Cohan, D., et al., 111
Cohen, L. J., 210
commons dilemma, 259
complementary events, 44
complete ordering axiom, 76
compound lottery, 77
  axiom, 78–9
conditional probability, 44–5, 153–4
conditional sampling, 156

Condorcet's paradox, 248
conservatism, 198, 215
continuity axiom, 77
continuous probability distributions,
    49–50, 109–11
correlation coefficient, 157
correlation, illusory, 153
Corrigan, B., 291
cost/benefit analysis, 208
credence decomposition, 129
cumulative distribution function
    (CDF), 51, 141–3, 228

Davis, D. L., 211
Dawes, R. M., 291, 293
De Groot, M. H., 246
de Neufville, R., 86
decidability axiom, 27
decision analysis
    partial, 4
    phases of, 114
    role of, 3–5, 7–8
Decision and Designs Inc., 255, 278–9
decision conferencing, 255
decision node, 66, 103
decision problems, features of, 2–3
decision structure, 112
decision support systems, 206
decision tables, 61
decision trees, 66, 102–22, 286
    assessment of, 112–22
    and continuous distributions, 109–11
    practical applications of, 111–12
Delphi, 253
dependence relationships, modeling,
    152–7
dependent events, 45
direct assessment of probabilities, 226
direct rating, 14
discrete probability distributions, 49,
    51–2, 109–11
disjoint events, 41–2
dogmatism, 201–2
dominance, 23
    stochastic, 141–3
Driver, M. J., 203
Drury, C., 146
Ducharme, W. M., 210
Duda, R. O., 284

ecological validity, 215
econometrics, 217, 221
Edwards, W., 10, 14, 19, 24, 83, 192,
    198, 236, 247
efficient deals, 276–7
efficient frontier, 23–5, 270–1, 276–8
Eilon, S., 157
Einhorn, H. J., 222, 234, 235, 296
EQUITY, 260, 271–3, 302
Ertel, S., 210
Esser, J. K., 252
event trees, 236–7
events, definition of, 38
exhaustive events, 42
expected monetary value (EMV)
    criterion, 61–5, 73
expected utility, 68–71
expected value, 52–3
expected value of imperfect
    information (EVII), 179–86
expected value of perfect information
    (EVPI), 177–9, 184
expert knowledge, 283, 293
expert systems, 281
expertise, 197, 219, 234, 281, 293
extended Pearson–Tukey
    approximation, 109–11

Farquahar, P. H., 79
Fatseas, V., 219
fault trees, 117, 236, 238
Fawkes, T. R., 157
Ferrell, W. R., 243, 246, 249, 255
finite upper and lower bounds axiom,
    27
Fischoff, B., 115, 117, 222, 235
forecasting methods, 212
forecasting practice, 218, 292
framing effects, 207, 208
French, S., 78, 251
Frenkel-Brunswik, E., 201

gambles, 207
Glendinning, R., 223
Goldberg, L. R., 291, 292
graph drawing in probability
    assessment, 229–30
group processes, 252

groupthink, 252, 256
Guerts, M. D., 223

Hadar, J., 164
Hayes-Roth, R., 284
Hershey, J. C., et al., 80–1
Hertz, D. B., 112, 138, 144, 156
Hespos, R. F., 151
heuristics, 193, 284
    anchoring and adjustment, 194
    availability, 194
    misperception of regression, 195
    representativeness, 194
hindsight, 235
HIVIEW, 31, 302
Hoerl, A., 222
Hogarth, R. M., 213
Hosseini, J., 112
Howard, R. A., 103, 128
Huber, G., 206
Hull, J. C., 157
Humphreys, P., 115

illusory correlation, 153
imperfect information, expected value
    of, 179–86
independent events, 45
influence diagrams, 103, 118–22
information, value of, 175–86
interval scales, 15, 263
investment decisions, 146–52
Irland, L. C., 223

Janis, I. R., 252
Jenks, J. M., 223
Johnson, E. J., 81, 297
joint probabilities, 46–7
Jones, J. M., 107
judgmental adjustments, 218, 294

Kabus, I., 222
Kahneman, D., 81, 153, 193, 194, 195,
    196, 207, 208, 232
Keefer, D. L., 109, 110
Keeney, R. L., 4, 8, 12, 86, 94, 115
Kiangi, G., 138
Kirkwood, C. W., 86

Klein, H. E., 223
knowledge elicitation, 287
knowledge engineers, 284
Kogan, N., 211
Kolmogoroff's axioms, 54

laboratory research, 192, 213, 219, 252,
    255, 293
Lacava, G. J., 185
Lawrence, M., 219, 220
Lawson, R. W., 216
Libby, R., 292
Lichtenstein, S., 194, 207, 222, 240
life underwriting, 287
Lindley, D., 233
Lindoerfer, J. S., 252
Lindstone, H., 258
linear models, 289
Lock, A., 246
log-odds scale, 239

Madden, T. J. et al., 112
Makridakis, S., 213
Malley, D. D., 223
marginal probability, 44–5
market research, 185–6
Markowitz, H. M., 143, 146
Marks, D. F., 240
mathematical aggregation of group
    judgments, 243–52
Matheson, J., 240
McCartt, A., 257
McGhee, W., 211
McNees, S. K., 223
mean–standard deviation screening
    procedure, 143–6
Meehl, P. E., 289, 293
Mock, T. J., 203
Monte Carlo simulation, 130–3
Moore, P. G., 37
Morgenstern, O., 65
Morris, P. A., 247
multi-attribute utility, 86–94
multi-attribute value, 19–29
multiple objectives, decisions
    involving, 2–3, 7–32, 86–95,
    259–79
multiplication rule of probability, 46–7
Murphy, A., 199, 215

mutual preference independence, 27–9
mutual utility independence, 87–8
mutually exclusive events, 41–2
Myers, I. B., 211
Myers–Briggs type indicator, 204

Nazareth, D. L., 297
negotiation models, 274–9
net present value (NPV)
  critique of, 148, 151–2
  introduction to, 146–9
  and simulation, 149–51
Newman, J. R., 14, 24
normal distribution, 144–5
normalization of weights, 20–1

objective, definition of, 8
odds, 37–8, 226, 239
Oliver, R. M., 128
outcome, definition of, 38
overconfidence, 214

package, definition of, 263
Pankoff, L. D., 223
Parenté, F. J., 258
Payne, J., 208
Pearson, E. S., 109
perfect information, expected value of,
  177–9, 184
personality, 201
  dogmatism, 201, 202
  risk taking, 202
Peterson, C. R., 210, 255, 278, 279
Phillips, L. D., 4, 30, 201, 202, 205,
  240, 255, 256, 259, 263, 279
Pitz, G., 196, 207, 208
Plax, T. G., 211
poker chips, 198
policy, definition of, 105
positiveness axiom, 54
posterior analysis, 174–5
posterior probability, definition of, 166
PREDICT! 156–7, 158, 302
preference independence, 27–9
preference orderings, aggregating for
  a group, 248–50
preposterior analysis, 175–86
prior probability, definition of, 166

probability
  aggregating group judgments of,
    243–4, 246–7
  assessment, 224–39
  classical, 39
  conditional, 44–5, 153–4
  density function (PDF), 50
  distribution, 48–53
  joint, 46–7
  marginal, 44–5
  posterior, 166
  prior, 166
  relative frequency, 39–40
  subjective, 40–1, 231–5
probability-equivalence approach, 67–9
probability method (for eliciting
    probability distributions), 228
probability theory
  axioms of, 54
  introductions to, 37–55
probability trees, 47–8, 236–7
probability wheel, 226–7
production systems, 197, 208, 284
prospect theory, 207

quadratic utility functions, 144–6

Raiffa, H., 8, 12, 83, 86, 94, 274, 278
random numbers, 131
rare events, assessment of probabilities
    for, 235–9
rationality, assumption of, 5, 85
real world, 200, 212, 215, 219, 293
Reinmuth, J. E., 223
relative frequency approach to
    probability, 39–40
relative heights, method of, 229–30
representativeness heuristic, 194
requisite decision models, 30–1
resource allocation problems, 259–74
  stages in modeling, 274
@RISK, 157, 158, 302
risk analysis, 129–58
risk aversion, 72–4
risk neutrality, 73
Rohrbaugh, J., 257
rollback method, 105–7
Ronen, B., et al., 76
Rothe, J. T., 223

Rowe, G., 258
Russell, W. R., 164

Scanlon, S., 185
scatter-diagrams, 154–6
Schell, G. P., 185
Schkade, D. A., 81
Schoemaker, P., 203
Seaver, D. A., et al., 230, 247
Selling, T. T., 293
sensitivity analysis, 25–6, 62–3, 135–6,
    184, 225, 273, 294
Shepanski, A., 292
Shortcliffe, E. H., 284
simulation, 129–58
  application to investment decisions,
      149–52
  stages in application to decision
      problems, 134
Singh, G., 138
Slovic, P., 83, 195, 207
SMART, 10
  axioms of, 26–7
Smedslund, J., 154
Soergel, R. F., 223
solvability axiom, 27
Spetzler, C. S., 184, 224, 230
St Petersburg paradox, 64
Stäel von Holstein, C. A., 224, 230, 240
standard deviation, 140, 143–6, 163–4
statistical extrapolation, 216, 219
statistical models of judgment, 289
Stigum, B. P., 100
stochastic decision trees, 151
stochastic dominance, 141–3
Strassman, P. A., 151
subjective probability, 40–1
  axioms, 231
  calibration, 202, 213, 233, 235
  coherence, 231–4
  consistency, 231–4
  reliability, 231
  validity, 233, 235
substitution axiom, 77
summation axiom, 77
SUPERTREE, 107, 302
Svenson, O., 207

Taylor, R. N., 210

Thaler, R. H., 81
Thomas, H., 37, 112, 138, 144, 230,
    239
Tocher, K. D., 83
transitivity, 27, 76, 248
Tukey, J. W., 109
Tull, D. S., 185
Tversky, A., 81, 83, 153, 193, 194, 195,
    196, 207, 208, 232

Ulvila, J. W., 111
uncertain quantity, 49
unequal probability axiom, 77–8
union of events, 42
  axiom, 54
utility
  as distinct from value, 9, 82–3
  axioms of, 76–9
  critique of, 82–5
  elicitation of, 65–9, 79–82
  expected, 68–71
  functions, 65–76, 79–83, 86–94
  for non-monetary attributes, 74–6
  independence, 87–8
  interpersonal comparisons of, 250–2
  interpretation of, 70–1, 72–4, 93–4
  multi-attribute, 86–94
  quadratic function, 144–6
  single attribute, 65–85

value trees, 11–13
value
  as distinct from utility, 9, 82–3
  assessment of, 14, 16–18, 263–5
  functions, 16–18, 275–6
  interpersonal comparisons of, 250–2,
      278
variables (in resource allocation
    problems), 261
von Winterfeldt, D., 19, 83, 115, 236,
    247
Von Neumann, J., 65
voting systems, 248–9

Wallsten, T. S., 230
Watson, S. R., 10, 13, 94
weather forecasting, 214, 234

weights
  across-criteria, 267–9
  importance, 19
  in averaging group judgments,
    245–6, 247
  swing, 19–21
  within-criterion, 265–7
Wenstop, F., 100
Weston, J. F., 146
Whitred, G., 292
Wilkins, D. C., 283
Willems, E. P., 223
Winkler, R., 199, 215, 240, 247

Winter, F. W., 112
within-criterion weights, 265–7
Wooler, S., 11, 31
word problems, 196
Wright, G., 7, 201, 202, 205, 222, 232,
  240, 294, 296

Youssef, Z. I., 210

Zamora, R. M., 184
Zimmer, I., 292